THE BIOLOGY BOOK

WORKBOOK

UNITS 3 4

Pam Borger
Kelli Grant
Louise Munro
Jane Wright

The Biology Book Units 3 & 4
1st Edition
Pam Borger
Kelli Grant
Louise Munro
Jane Wright

Contributing authors: Daniel Avano, Andrea Blunden, Sue Farr, Sarah Jones and Katrina Walker

Publisher: Rachel Ford
Project editors: Nadine Anderson-Conklin and Siobhan Moran
Permissions researcher: Lyahna Spencer
Cover image: iStock.com/nopparit
Production controller: Karen Young
Typeset by: MPS Limited

Any URLs contained in this publication were checked for currency during the production process. Note, however, that the publisher cannot vouch for the ongoing currency of URLs.

For product information and technology assistance,
in Australia call **1300 790 853**;
in New Zealand call **0800 449 725**

For permission to use material from this text or product, please email **aust.permissions@cengage.com**

ISBN 978 0 17 041174 5

Cengage Learning Australia
Level 7, 80 Dorcas Street
South Melbourne, Victoria Australia 3205

Cengage Learning New Zealand
Unit 4B Rosedale Office Park
331 Rosedale Road, Albany, North Shore 0632, NZ

For learning solutions, visit **cengage.com.au**

Printed in China by 1010 Printing International Limited.
2 3 4 5 6 7 23

CONTENTS

TOPIC 2: ECOSYSTEM DYNAMICS

9780170411745

UNIT 4 » HEREDITY AND CONTINUITY OF LIFE 56

TOPIC 1: DNA, GENES AND CONTINUITY OF LIFE

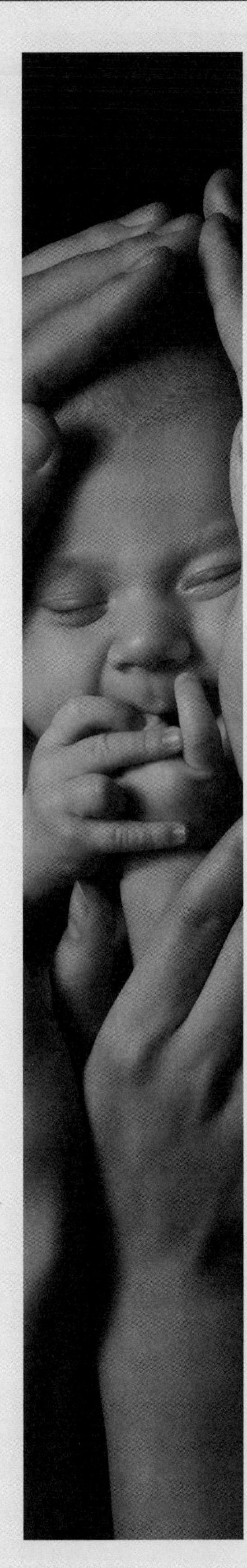

TOPIC 2: CONTINUITY OF LIFE ON EARTH

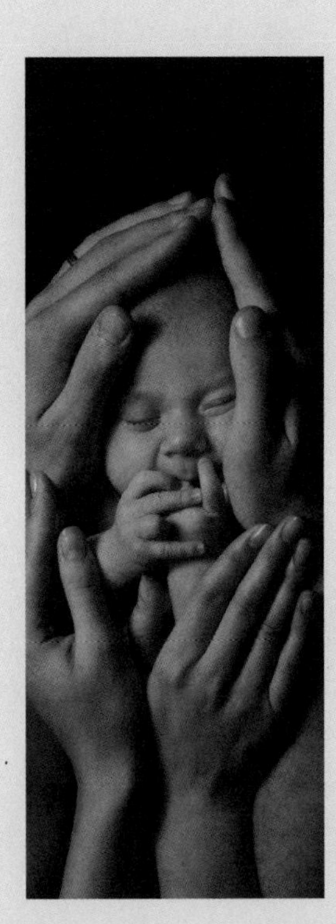

HOW TO USE THIS BOOK

Learning

The learning section is a summary of the key knowledge and skills. This summary can be used to create mind maps, to write short summaries and as a check list.

Revision

This section is a series of structured activities to help consolidate the knowledge and skills acquired in class.

Evaluation

The evaluation section is in the style of a practice exam to test and evaluate the acquisition of knowledge and skills.

Practice exam

A tear-out exam helps to facilitate preparing and practicing for external exams.

ABOUT THE AUTHORS

Pam Borger

Pam is a highly experienced biology teacher and author. Among Pam's writing credits is the successful *Jumpstart Biology*. Pam has been instrumental in delivering and organising professional development for Biology teachers.

Kelli Grant

Kelli has a lot of experience teaching biology, with breadth across the high school, TAFE and tertiary sectors.

Louise Munro

Louise is a highly experienced teacher of senior biology and chemistry. Her areas of expertise include molecular biology and biochemistry.

Jane Wright

Jane is an experienced biology teacher and writer. Among Jane's writing credits is *Nelson Biology for the Australian Curriculum*.

9780170411745

SYLLABUS REFERENCE GRID

UNITS AND TOPICS	THE BIOLOGY BOOK UNITS 3 & 4
UNIT 3: BIODIVERSITY AND THE INTERCONNECTEDNESS OF LIFE	
Topic 1: Describing biodiversity	
Biodiversity	Chapter 1
Classification process	Chapter 2
Topic 2: Ecosystem dynamics	
Functioning ecosystems	Chapter 3
Population ecology	Chapter 4
Changing ecosystems	Chapter 5
UNIT 4: HEREDITY AND CONTINUITY OF LIFE	
Topic 1: DNA, genes and continuity of life	
DNA structure and replication	Chapter 6
Cellular replication and variation	Chapter 7
Gene expression	Chapter 8
Mutations	Chapter 9
Inheritance	Chapter 10
Biotechnology	Chapter 11
Topic 2: Continuity of life on Earth	
Evolution	Chapter 12
Natural selection and microevolution	Chapter 13
Speciation and macroevolution	Chapter 14

UNIT THREE

BIODIVERSITY AND THE INTERCONNECTEDNESS OF LIFE

- Topic 1: Describing biodiversity
- Topic 2: Ecosystem dynamics

Alamy Stock Photo/Siegfried Schnepf

1 Biodiversity

LEARNING

Summary

- Biodiversity includes the diversity of species and ecosystems.
- The diversity of species can be determined using indices such as:
  - species richness, the number of species present in an ecosystem;
  - relative species abundance, the number of individuals present for each species in an ecosystem;
  - percentage cover, the percentage of the quadrat that a species takes up;
  - percentage frequency, the percentage of quadrats that a species appears in; and
  - Simpson's Diversity Index, the combined ratio of individuals in each species to the total individuals in an ecosystem.
- Simpson's Diversity Index is calculated with the formula $D = 1 - \left(\frac{\Sigma n(n-1)}{N(N-1)}\right)$ where n is the number of individuals of each species and N is the total number of individuals at the site.
- Ecosystems can be compared across spatial (area) scales, such as micro-, meso- and macro-level ecosystems.
- Ecosystems can be compared across temporal (time) scales, such as days, seasons or years.
- Environmental factors, such as species diversity, species interactions and abiotic factors, limit the distribution and abundance of species in an ecosystem.

9780170411745

1.1 The Great Barrier Reef: biodiversity hotspot

The World Heritage Committee meets once a year and consists of 21 representatives elected from 190 member countries. One of the committee's tasks is to recommend inscriptions (inclusions) on the World Heritage List. The List aims to recognise and protect sites of cultural and natural significance.

The Great Barrier Reef is a site of remarkable variety and beauty on the north-east coast of Australia. It contains the world's largest collection of coral reefs. It also holds great scientific interest as it is the habitat of species such as the dugong and the large green turtle, both of which are threatened with extinction.

As the world's most extensive coral reef ecosystem, the Great Barrier Reef is a globally outstanding and significant entity. Practically the entire ecosystem was inscribed on the World Heritage List in 1981, covering an area of 348 000 km^2. The Great Barrier Reef includes extensive cross-shelf diversity, stretching from the low water mark along the mainland coast up to 250 km offshore. This wide depth range includes vast shallow inshore areas, mid-shelf and outer reefs, and beyond the continental shelf to oceanic waters more than 2000 m deep.

Within the Great Barrier Reef there are some 2500 individual reefs of varying sizes and shapes, and more than 900 islands, ranging from small sandy cays (small island) and larger vegetated cays, to large rugged continental islands rising, in one instance, to more than 1100 m above sea level. Collectively these landscapes and seascapes provide some of the most spectacular maritime scenery in the world.

On many of the cays, there are spectacular and globally important breeding colonies of seabirds and marine turtles; Raine Island is the world's largest green turtle breeding area. On some continental islands, large aggregations of over-wintering butterflies periodically occur.

The latitudinal and cross-shelf diversity, combined with diversity through the depths of the water column, encompasses a globally unique array of ecological communities, habitats and species. This diversity of species and habitats, and their interconnectivity, make the Great Barrier Reef one of the richest and most complex natural ecosystems on earth. There are more than 1500 species of fish, approximately 400 species of coral, 4000 species of mollusk, some 240 species of birds, plus a great diversity of sponges, anemones, marine worms, crustaceans and other species. No other World Heritage site contains such biodiversity. This diversity, especially the endemic species, means the Great Barrier Reef is of enormous scientific and intrinsic importance, and it also contains a significant number of threatened species.

Source: UNESCO/CLT/WHC, from http://whc.unesco.org/en/list/154, Great Barrier Reef article

QUESTIONS

1 Define biodiversity in the context of the Great Barrier Reef.

___

___

2 The *International Union for the Conservation of Nature* evaluation stated '... if only one coral reef site in the world were to be chosen for the World Heritage List, the Great Barrier Reef is the site to be chosen'. Give three reasons why the Great Barrier Reef is an exemplary coral reef.

___

___

___

___

3 One of the selection criteria for inclusion on the World Heritage List is 'to contain the most important and significant natural habitats for *in situ* conservation of biological diversity, including those containing threatened species of outstanding universal value from the point of view of science or conservation.' How does the Great Barrier Reef meet this criterion?

4 Evaluate the goals and efficacy of declaring natural areas World Heritage sites.

1.2 The five measures of species diversity

1 Complete the table to summarise the five measures of species diversity.

MEASURE OF SPECIES DIVERSITY	DEFINITION	EXAMPLE
Species richness		
Relative species abundance		
Percentage cover		
Percentage frequency		
Simpson's Diversity Index (D)		

1.3 Simpson's Diversity Index

Simpson's Diversity Index gives a number between zero (no diversity) and one (infinite diversity), while accounting for both species richness and relative species abundance.
It is calculated with this formula:

$$D = 1 - \left(\frac{\sum n(n-1)}{N(N-1)} \right)$$

where n is the number of individuals of each species and N is the total number of individuals at the site.

STIMULUS DATA				
SPECIES	INDIVIDUALS AT SITE 1	INDIVIDUALS AT SITE 2	INDIVIDUALS AT SITE 3	INDIVIDUALS AT SITE 4
Banksia sp.	1	2	0	3
Callistemon sp.	2	3	1	3
Xanthostemon verticillatus	6	7	5	4
Bouteloua dactyloides	12	9	30	6
Myoporum parvifolium	9	9	6	2

QUESTIONS

1 Calculate the value of D for each site.

2 Arrange the sites in order from least biodiverse to most biodiverse. Comment on whether this order is as you would have predicted.

3 Site 4 has considerably less life than Site 3 and yet, Simpson's Diversity Index shows it to be more biodiverse. Explain why this is so.

4 Compare the values for D at Sites 2 and 4.

5 A disease that affects *Bouteloua* species is introduced to the four sites, killing all individuals. Comment on the impact this has on the biodiversity of each site.

1.4 Spatial and temporal comparisons

1 Complete the table by giving examples of how species diversity differs in each case.

TEMPORAL SCALE	AT DIFFERENT TIMES OF DAY	AT DIFFERENT TIMES OF YEAR	AT DIFFERENT TIMES OVER A DECADE
How species diversity may differ in the same place:			
SPATIAL SCALE	**IN DIFFERENT LOCAL AREAS**	**IN DIFFERENT STATES**	**ON DIFFERENT CONTINENTS**
How species diversity may differ at the same time:			

9780170411745

1.5 Environmental limiting factors

1 Create a mindmap to show the connections between the words in the wordbank. Include relevant information about each of the words and how they affect each other.

Population size	Predation	Temperature
Rainfall	Shelter	Interspecies competition
Food source	Soil conditions	Intraspecies competition

1 Biodiversity is:

A the number of species present in an ecosystem.

B a group of organisms that share a gene pool.

C the full range of living things in a particular area.

D the interactions between a community and the environment.

2 Which of the following formulae will accurately calculate Simpson's Diversity Index?

A $D=1-\left(\frac{\Sigma n(n-1)}{N(N-1)}\right)$

B $D=1-\left(\frac{\Sigma N(N-1)}{n(n-1)}\right)$

C $D=1-\left(\frac{\Sigma n(n-1)}{(N-1)}\right)$

D $D=1-\left(\frac{\Sigma n(N-1)}{N(n-1)}\right)$

3 List one biotic and one abiotic factor that limit species abundance and distribution.

____

____

4 State which of the five measures of species biodiversity (you may choose more than one) you would use to describe the biodiversity of:

a a coral reef

____

b a grassy meadow

____

c a tidal rock pool

____

5 Explain how the measured biodiversity of an urban street changes between daytime and night time hours.

____

____

____

6 A student claims in their Student Experiment that the biodiversity of the local woodland is greater than that of their backyard. Their averaged data is displayed in Table 1.6.1. Analyse the data to determine if it supports or rejects their claim. Communicate your findings to the student in a professional manner.

TABLE 1.6.1

LOCAL WOODLAND SITE	AVERAGE NUMBER OF INDIVIDUALS
SPECIES	
Acacia sp.	4
Astrotricha sp.	5
Cryptostylus sp.	1
Eucalyptus sp.	6
Harpullia pendula	1
Scaevola aemula	2
BACKYARD SITE	AVERAGE NUMBER OF INDIVIDUALS
SPECIES	
Brachyscome multifida	6
Pandorea jasminoides	3
Westringia sp.	10
Elaeocarpus reticulatus	1
Leptospermum petersonii	2

Analyse:

Communicate:

2 Classification process

LEARNING

Summary

- Biological classification can be hierarchical and based on different levels of similarity of physical features, methods of reproduction and molecular sequences.
- The Linnaean system generally classifies organisms based on similarity of physical features. Modifications to the original Linnaean system include reproductive and molecular similarities.
- K/r selection is a subset of reproductive classification that separates species that reproduce quickly, in large numbers and without intensive parenting, from those that reproduce slowly, in small numbers and with considerable parental investment.
- Cladistics is a system that classifies organisms based on their molecular similarity, including DNA and proteins. The organising unit of cladistics is the clade: a group of organisms comprising all of the descendants of a particular ancestor organism.
- The three underlying assumptions of cladistics are
  - common ancestry: as all life evolved from a single ancestor, any group of organisms will share a common ancestor at some point in the past;
  - dichotomous cladogenesis: the offspring of an ancestral species diverge dichotomously in a process called cladogenesis; and
  - physical change: organisms become increasingly different as they continue evolving from their point of cladogenesis.
- Evolutionary relatedness can be inferred from cladograms and from direct molecular sequences. Increasing similarity denotes increasing relatedness.
- Multiple definitions of species are required to accommodate different ways of determining relatedness, including the biological species concept, the morphological species concept and the phylogenetic species concept.
- Hybrids are the offspring of two different species and are usually infertile.
- Species interactions such as predation, competition, symbiosis and disease can also be used to classify organisms.
- Ecosystems can be classified according to the various habitats they are composed of, from microhabitats up to ecoregions. Specht's ecosystem classification system is used in Australia to group similar ecosystems for combined study.
- Stratified sampling is often used to more accurately collect ecological data, such as determining population size, density and distribution, and involves careful site selection, use of quadrats and transects, and several methods to minimise bias.

2.1 Classification as a diagnostic tool

Read the following account of Dr Adam Jenney's work to answer the questions that follow.

Staphylococcus aureus or 'golden staph' is a bacterium that can cause serious infections of the skin, blood, bones and other organs. Some strains of *S. aureus* are resistant to commonly used antibiotics and need to be treated differently to other bacteria. It can be difficult to tell which bacterium is causing an infection because different bacteria can cause similar symptoms in patients. So how do doctors know what bacterium they need to treat?

Dr Adam Jenney is a clinical microbiologist at the Alfred Hospital in Melbourne. He works with other doctors and scientists to help them identify bacteria that are causing disease in patients. According to Dr Jenney, 'identifying organisms allows you to predict how the pathology associated with that infection may evolve'. Most importantly, accurate identification is necessary because 'we treat different species in different ways. An example of that is *S. aureus* versus other types of *Staphyloccocus*'.

One of the first steps in identifying these bacteria is to grow the organisms in the laboratory. 'Let's take, for example, a specimen of sputum from someone who the clinician thinks has got pneumonia ... we take a sterile swab and will put some of that sputum onto culture media.' The next step, describes Dr Jenney, is to 'spread the sputum over the plate to allow us to achieve, if there is something to grow, single colonies so we can identify the organisms causing the infection.'

Until recently, organisms were identified using classification keys based on physical and biochemical properties. 'We would take a tiny amount from a colony that we think might be a pathogen, spread it on a glass slide and make a Gram stain of it,' explains Dr Jenney. 'Gram stain allows us to fix colour into organisms to determine whether it is gram positive or negative, which is still very important in the identification process.' The results of the Gram stain help to direct further tests to be performed. 'For instance, if the slide showed gram-positive cocci [spherical-shaped bacteria], we could take another small dab of the colony and mix it with some hydrogen peroxide on a clean glass slide. If this produced bubbles, it indicates oxygen is being released. This is a positive catalase test indicative of staphylococci rather than the other common gram-positive cocci we see, namely streptococci.'

New technologies, however, have changed the way that bacteria are identified. The laboratory at the Alfred now uses a MALDI-TOF mass spectrometer to identify bacterial colonies. 'You take a normal, though sterile, toothpick and pick off a little bit of that colony and place it on a slide,' outlines Dr Jenney. 'Then the MALDI-TOF fires lasers at the colony forcing proteins to be released. Those proteins have a characteristic size and they are able to be identified using a mass spectrometer.' Mass spectrometry measures the size of particles by converting them to ions and separating them by size and charge. Each bacterial species is made up of a different set of proteins, so a computer can use this information to identify the species of bacteria. The MALDI-TOF can quickly diagnose a *S. aureus* infection, the significance of which is often considerably greater than other species of staphylococcus.

As well as helping with identification, Dr Jenney also works with organisations to help track and prevent the spread of infections within hospitals and the community. He became interested in microbiology during his training as a doctor specialising in infectious diseases. 'I realised that such a vast amount of clinical infectious diseases relies on clinical microbiology so I did that as well.' Dr Jenney's advice for science students considering a career in science: 'Do it! Science is forever expanding and is becoming more interesting and is constantly challenging.'

QUESTIONS

1 Explain why it is important for doctors to be able to identify species of bacterium.

2 Construct a flow chart to show how bacteria are identified in the laboratory.

3 Distinguish between the characteristics of bacteria used in identification through traditional microbiological techniques, such as growing them on a plate, and MALDI-TOF mass spectrometry.

4 Explain how collaboration between scientists from different fields has changed the way that bacterial identification occurs.

2.2 Linnaean classification

1 Complete the table to summarise the levels of the Linnaean classification system.

LEVELS OF CLASSIFICATION	EXAMPLES
Kingdom	Animalia (animals), Plantae (plants) etc.

2 The Linnaean system has been modified over the years to include reproductive and molecular homologies. Give two examples of organisms that have been reclassified due to improved understandings of relatedness.

9780170411745

2.3 K/r selection

K/r selection stems from the statistical analysis of populations, where this ratio represents the trade-off between quantity and quality of offspring. In a grossly oversimplified manner, K represents parental involvement in raising offspring, while r represents the number of offspring produced each year.

Species who are considered to operate under K selection produce one or two offspring each year (small r) so that they can put a lot of time and effort into feeding, protecting and raising them (large K). This may result in fewer offspring in total, but a higher percentage surviving to adulthood.

Species who are considered to operate under r selection can produce upwards of a dozen offspring each year (large r) but provide limited parental support in feeding, protecting or raising them (small K). This may result in considerably more offspring in total, but a much lower percentage survive to adulthood.

Both methods are effective in the continuation of the species, but in some cases, one will be more advantageous than the other.

QUESTIONS

1 Sort the following organisms according to their K/r selection.

Whales	Mice	Plants	Elephants	Humans	Dogs	Magpies
Ants	Locusts	Turtles	Crocodiles	Salmon	Coral	

K SELECTION	R SELECTION

2 Some of the organisms from question 1 will have been difficult to place. Explain why these organisms don't appear to fall into a neat category.

____

____

____

3 Suggest a modification for this system that would accommodate the difficult organisms.

____

____

____

2.4 Cladistics

Cladistics is the name given to the process of organising species into clades and constructing cladograms. Character matrices, tables that collate present or absent traits, can be very useful in this process.

QUESTIONS

Construct a cladogram of the organisms in Figure 2.4.1.

FIGURE 2.4.1 The starfish, snail, bee, cow, duck and turtle are all related in some way.

1 Complete a character matrix of the organisms from Figure 2.4.1 to show which characteristics are present or absent. You will need to identify several more characteristics to highlight the differences between them all.

	STARFISH	SNAIL	BEE	COW	DUCK	TURTLE
Bilateral symmetry						
Has a backbone						
Gives birth to live young						

2 Use a separate piece of paper to draft a cladogram. Select the two organisms that share the most characteristics and place these as an initial branch on your cladogram.

3 Arrange the rest of the organisms such that the cladogram is as simple as possible.

4 Check your cladogram by marking which characteristic evolved at each node and reshuffling the organisms to eliminate as many 'double-ups' as possible.

5 Copy your completed cladogram into the space here.

6 Circle a clade on your cladogram.

7 If there were multiple ways to arrange your organisms, explain why you chose your particular construction.

8 From your experience with this activity, summarise what a cladogram represents.

____

____

____

____

2.5 Underlying assumptions of cladistics

1 Complete the table to describe the three main assumptions that underpin the field of cladistics.

ASSUMPTION	DESCRIPTION	EXAMPLE
Common ancestry	____________	____________
____________	The offspring of an ancestral species diverge dichotomously in a process called cladogenesis.	____________
____________	____________	Organisms that are closely related, such as chimpanzees and humans, share many similarities, while organisms that are more distantly related, such as humans and snails, share comparatively fewer similarities.

2.6 Determining evolutionary relatedness

1 Construct a cladogram to show probable evolutionary relatedness between the eight phyla of Kingdom Animalia: Porifera, Cnidaria, Echinodermata, Arthropoda, Annelida, Mollusca, Platyhelminthes and Chordata. The dichotomous key in Figure 2.6.1 may be helpful in identifying the characteristics of each class.

1a	Body symmetrical	go to 2
1b	Body asymmetrical with no structure/organs	Phylum Porifera
2a	Body with multiple planes of symmetry (radial symmetry)	go to 3
2b	Body with a single plane of symmetry (bilateral symmetry)	go to 4
3a	Tentacles present, body soft	Phylum Cnidaria
3b	No tentacles, body hard and rough	Phylum Echinodermata
4a	Body divided up into segments (segmented)	go to 5
4b	Body unsegmented	go to 6
5a	Body divided into 2–3 segments; jointed legs present	Phylum Arthropoda
5b	Body divided into many segments; jointed legs absent	Phylum Annelida
6a	Body with internal or external shell	Phylum Mollusca
6b	Body without shell	go to 7
7a	Body flat and worm-like	Phylum Platyhelminthes
7b	Body not worm-like	Phylum Chordata

FIGURE 2.6.1 The eight phyla of Kingdom Animalia have many distinct differences.

2 Construct a cladogram to show probable evolutionary relatedness between the seven classes of Phylum Chordata: Mammalia, Aves, Osteichthyes, Chondrichthyes, Agnatha, Reptilia and Amphibia. The dichotomous key in Figure 2.6.2 may be helpful in identifying the characteristics of each class.

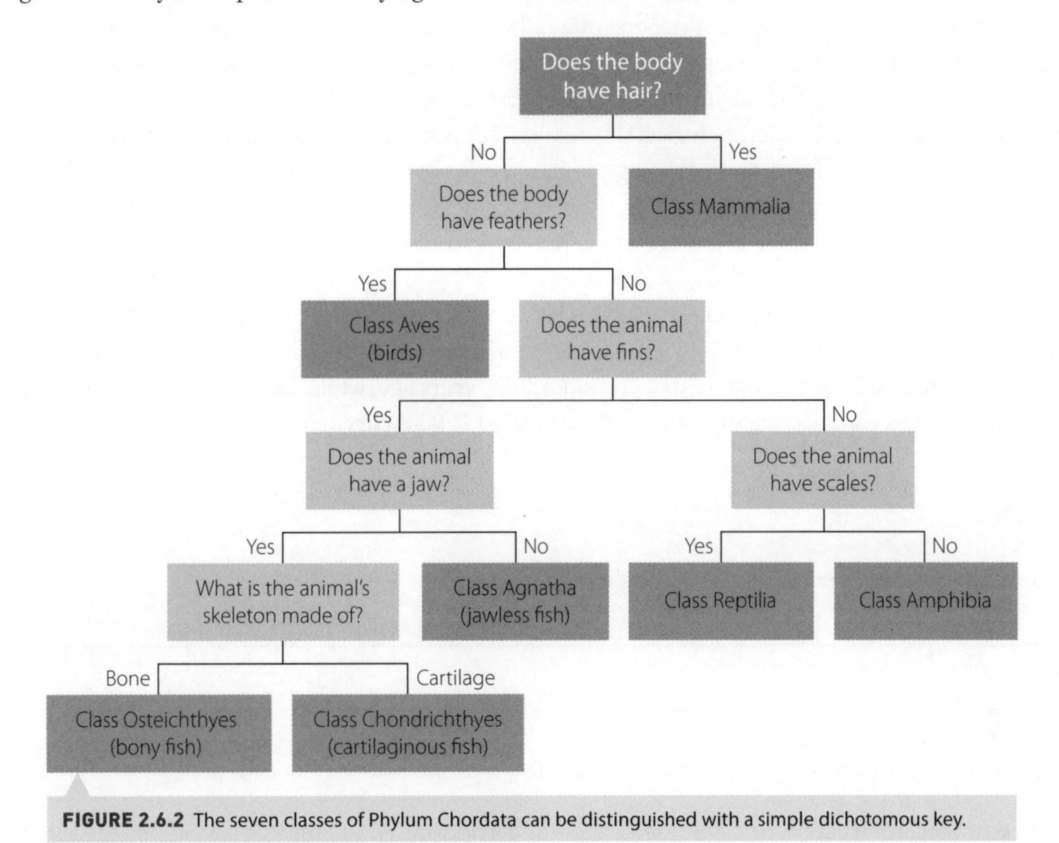

FIGURE 2.6.2 The seven classes of Phylum Chordata can be distinguished with a simple dichotomous key.

2.7 Multiple definitions of species

In biological terms, a species is a group of organisms whose members have the potential to interbreed in nature and produce viable, fertile offspring. Most importantly, individuals within a species are reproductively isolated from individuals not belonging to that species. This biological species concept is the predominant definition of a species, however, it requires organisms to be living to determine reproductive compatibility.

When examining fossils, the morphological species concept is most commonly applied. This concept characterises a species by its form, or morphology. In the case of the human family tree, it is most often the skull that is best preserved and identified morphologically. However, there are many disagreements between scientists about which morphological features should be used, and when the features are sufficiently different to justify the creation of a new group.

The phylogenetic species concept identifies a species as being the smallest clade possible, or smallest group of organisms who can all trace their origins to a single common ancestor. For example, all members of the species *Homo sapiens* share a common ancestor that existed around 200 000 years ago, while domesticated dogs of the species *Canis familiaris* share a common ancestor that existed around 15 000 years ago.

QUESTIONS

1 Define the three species concepts with a single sentence each.

____

____

____

2 Describe how each concept would consider a horse and a donkey. Would they be considered the same or separate species? What reasoning would be used?

____

____

____

____

____

3 Suggest two limitations and a benefit of having multiple definitions of species.

____

____

____

2.8 Hybrids

Often when two species mate, their offspring are no more or less suited for survival and they pass away after their lifespan, taking their hybrid genes with them. However, some hybrid species feature the best strengths of their parents and are better suited to human use than either of their parent species. This is the case with horse-donkey hybrids called mules.

Mules are produced when a female horse is impregnated by a male donkey. Female donkeys impregnated by male horses produce hinnies, but are harder to bring to term. Mules have the size and carrying capacity of a horse, but are considerably stronger and more resilient. They have the endurance and tenacity of a donkey but are considerably more affectionate and easier to train. They require less food and less attention, and are more resistant to sunburn, insect bites, injury and equine disease than either of their parents. In short, mules are a super-equine, that are unfortunately infertile.

There is evidence of mule breeding in ancient Greek times, which may explain why their name does not follow general hybrid conventions. In general, the first part of the name is taken from the Latin name of the sire (father), followed by part of the Latin name for the mother. This results in two names for the offspring of two species, depending on which species actually gave birth to the offspring. Ligers and tigons, for example, are both lion and tiger hybrids. The former had a tiger mother and the latter, a lion mother. If mules followed this convention, they would be named dorses.

QUESTIONS

1 Select two animals of different species and sketch an image of their hybrid offspring. Name this offspring using the general conventions and provide a short description of the features it has from each of its parents.

2.9 Species interactions

1 Categorise the following species interactions according to their relationships.

A snake eating a mouse
A ramora fish eating shark parasites
An owl stealing another owl's prey
A male kangaroo fighting another for females
A rabbit infected with the myxoma virus
A cane toad eating a frog
A clownfish living within a sea anemone
A koala infected with chlamydia
A tapeworm colony in the gut of a dog
A magpie swooping a cyclist
A feral cat eating a parakeet
A cattle egret riding on a cow to eat the insects disturbed by its feeding
A possum eating a eucalypt flower
Two colonies of ants fighting over a dead grasshopper
A Tasmanian devil with a Devil Facial Tumour
An epiphyte growing in the canopy of a tall tree

COMPETITION	PREDATION

SYMBIOSIS	DISEASE

9780170411745

2.10 Classifying ecosystems

There are many ways to classify ecosystems. The Holdridge life zone classification system, Specht's classification system, the ANAE classification system and EUNIS habitat classification system are just four of the systems used in Australia.

QUESTIONS

1 Outline the major features, benefits and drawbacks of each system.

CLASSIFICATION SYSTEM	FEATURES	BENEFITS	DRAWBACKS
Holdridge life zone			
Specht's			
ANAE			
EUNIS			

2 Comment on whether Australia should move to a single, standard classification system.

2.11 Stratified sampling

Use the following prompts to develop an informative essay on the use of stratified sampling in ecology. It will assist you if you respond to each prompt in complete paragraphs.

1 Define stratified sampling.

2 Describe the purpose of stratified sampling. Why do ecologists use it? What can the data be used for?

3 Outline the considerations of site selection that stratified sampling addresses. How does this process make site selection easier or broader? What makes a site ideal for stratified sampling?

4 Describe the surveying techniques, such as quadrats and transects, that are used in stratified sampling. When are they used?

5 Outline the techniques that ecologists use to minimise bias in their data. How do they ensure comparability and replicability?

6 Describe the main methods of data representation from stratified sampling. How do ecologists analyse their data? How do they communicate and corroborate their findings?

1 Which of the following is not a recognised definition of species?

A biological species concept

B ecological species concept

C morphological species concept

D phylogenetic species concept

2 In K/r selection, K selection refers to organisms that:

A produce many offspring with little parental involvement.

B produce few offspring with little parental involvement.

C produce few offspring with intense parental involvement.

D produce many offspring with intense parental involvement.

3 Name the eight major taxonomic levels in Linnaean classification.

____

____

4 Table 2.12.1 shows a simplified dichotomous key for identifying several species of bacteria that can cause disease in humans. Using this key involves performing a Gram stain, which is a staining process that groups bacteria into two groups based on characteristics of their cell wall.

TABLE 2.12.1 Identifying bacteria

1a	Gram positive (stains purple with a Gram stain)	go to 2
1b	Gram negative (stains pink with a Gram stain)	go to 4
2a	Bacteria are cocci (sphere-shaped)	go to 3
2b	Bacteria are bacilli (shaped like rods)	*Clostridium difficile*
3a	Cocci are arranged in chains	*Streptococcus pneumonia*
3b	Cocci are arranged in bunches	*Staphylococcus aureus*
4a	Bacteria are cocci (sphere-shaped)	*Neisseria* spp.
4b	Bacteria are bacilli (shaped like rods)	*Escherichia coli*

a State which of the listed species are bacilli.

____

b Using the dichotomous key, determine what species of bacteria is shown in Figure 2.12.1.

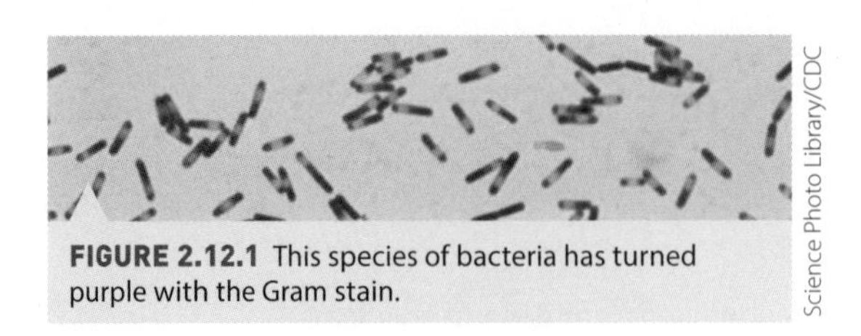

FIGURE 2.12.1 This species of bacteria has turned purple with the Gram stain.

c Not all bacteria can be identified to the species level with this key. Identify which group is unable to be specifically identified and provide a reason why this may be.

5 Given the two cladograms in Figure 2.12.2, determine which is more likely to be correct and justify your decision with cladistics.

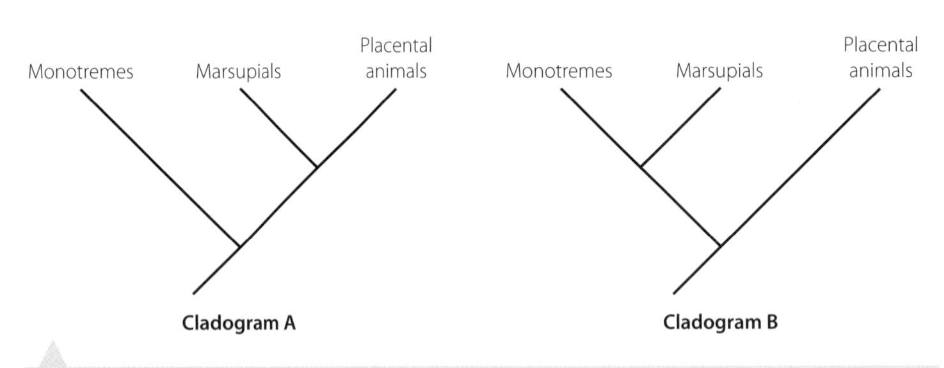

FIGURE 2.12.2 The evolution of marsupials has puzzled scientists for decades.

6 Advances in remote sensing radar imagery and satellite tracking in real time have enabled scientists to measure and monitor populations, and to play a significant role in surveying and monitoring large or inaccessible ecosystems. Construct an argument for or against the use of advanced technology in monitoring ecosystem trends.

9780170411745

3 Functioning ecosystems

LEARNING

Summary

- Photosynthesis transforms solar energy into chemical energy, which is then transferred through the food chain.
- Chemical energy is used by organisms to move, grow and reproduce. Growth and reproduction contribute to increasing biomass, the dry weight of organic matter in an ecosystem.
- Energy transfers through food chains, food webs and pyramids can be analysed and calculated by beginning with gross inputs and incorporating 10% transfer efficiency from one trophic level to the next.
- Energy is lost at each transfer through radiation and respiration.
- Energy flow diagrams (such as Figure 3.0.1) illustrate the movement of energy through an ecosystem, including the productivity (gross and net) of the various trophic levels.

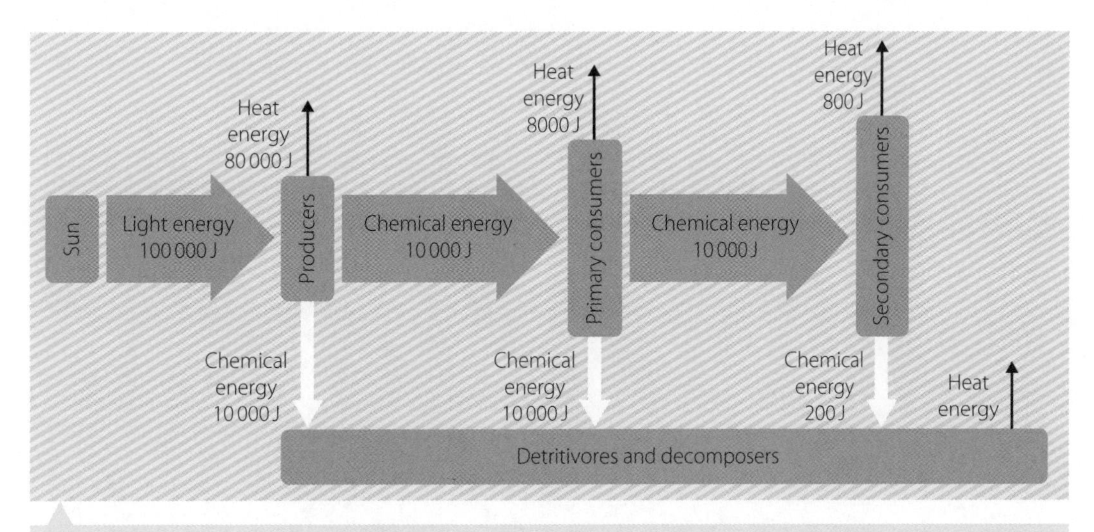

FIGURE 3.0.1 Energy flow diagrams follow the journey of the Sun's energy through an ecosystem.

- Matter also flows through ecosystems. The carbon, nitrogen and water cycles show this movement.
- An ecological niche is the role and space that an organism fills in an ecosystem, including all its interactions with the biotic and abiotic factors of its environment.
- The competitive exclusion principle holds that no two organisms can occupy the exact same niche for long. They will compete until the better-suited organism excludes the other from the niche.
- A keystone species is a plant or animal that plays a unique and crucial role in the way an ecosystem functions. The structure of their community is dependent upon their interactions.

3.1 Food chains

1 Construct a basic food chain and annotate it to include:

a energy transformation

b energy transfer

c energy loss

d interactions with the carbon cycle

e biomass production

3.2 Calculating energy transfers

$$Percentage\ efficiency = \frac{net\ productivity\ of\ the\ organism}{net\ productivity\ of\ the\ previous\ trophic\ level} \times 100$$

1 Calculate the energy available to the top-order predator in each of the following food chains. Assume 10% transfer efficiency.

a Sun → Producers – 500 kJ → 1st Consumers → 2nd Consumers → 3rd Consumers

b Sun – 1000 kJ → Producers → 1st Consumers → 2nd Consumers → 3rd Consumers → 4th Consumers

c Sun – 3000 kJ → Producers → 1st Consumers → 2nd Consumers

2 Calculate the transfer efficiency of the following food chains.

a Sun – 50 000 kJ → Grass – 500 kJ → Rabbit – 200 kJ → Snake – 10 kJ → Eagle

b Sun – 10 MJ → Eucalypt – 6000 kJ → Koala – 1000 kJ → Dingo

c Sun – 30 000 kJ → Tree – 3000 kJ → Caterpillar – 600 kJ → Bird – 50 kJ → Cat – 9 kJ → Crocodile

3 An open forest ecosystem with an even mix of grasses and eucalypt-like trees receives 10 million joules of energy from the Sun. This ecosystem is populated with small birds, insects, possums, rabbits and snakes.

a Construct a food web with assumed 10% energy transfer for this ecosystem.

b A medium-sized carnivore, such as a dingo, requires at least 5000 kJ per day to survive. Given the amount of energy available in the various food sources, calculate how many dingoes could survive here and comment on whether a pack would reasonably be able to thrive.

3.3 Energy flow diagrams

Energy flow diagrams (such as Figure 3.0.1) map the journey of the Sun's energy as it transfers through the trophic levels of an ecosystem and is transformed into heat by metabolism. Several key principles govern the construction of these diagrams.

- Arrows show the direction of energy flow and are labelled with the form and quantity of energy they carry.
- Boxes are used to represent trophic levels in sequence.
- Heat energy from metabolic processes is lost from each trophic level to the surroundings.
- Detritivores and decomposers are included as a subtrophic level, as they play an important role in energy transformation.
- Total heat energy lost must equal the amount of chemical energy brought into the system by producers.

QUESTIONS

1 Complete an energy flow diagram for each of the following food chains.

a Sun – 50000 kJ → Grass – 500 kJ → Rabbit – 200 kJ → Snake – 10 kJ → Eagle

b Sun – 10 MJ → Eucalypt – 6000 kJ → Koala – 1000 kJ → Dingo

c Sun – 30000 kJ → Tree – 3000 kJ → Caterpillar – 600 kJ → Bird – 50 kJ → Cat – 9 kJ → Crocodile

9780170411745

2 Explain two limitations of energy flow diagrams when trying to map the journey of the Sun's energy in an ecosystem.

3.4 The carbon cycle

The carbon cycle is unique among nutrient cycles because it does not necessarily involve decomposers. In the absence of any decomposers, carbon would still be able to circulate for some time within an ecosystem since carbon is incorporated into and released from glucose through photosynthesis and respiration.

QUESTIONS

1 Draw a detailed diagram of the carbon cycle, including the following terms: diffusion, weathering, respiration, photosynthesis, absorption and combustion.

3.5 The nitrogen cycle

The nitrogen cycle (Figure 3.5.1) is a combination of two cycles.

1 The elemental cycle, where N_2 is absorbed from the atmosphere by nitrogen-fixing bacteria and released back to the atmosphere by denitrifying bacteria and volcanic activity.

2 The ionic cycle, where nitrogen-containing ions such as nitrate (NO_3^-) and ammonium (NH_4^+) are passed between organisms in the biosphere.

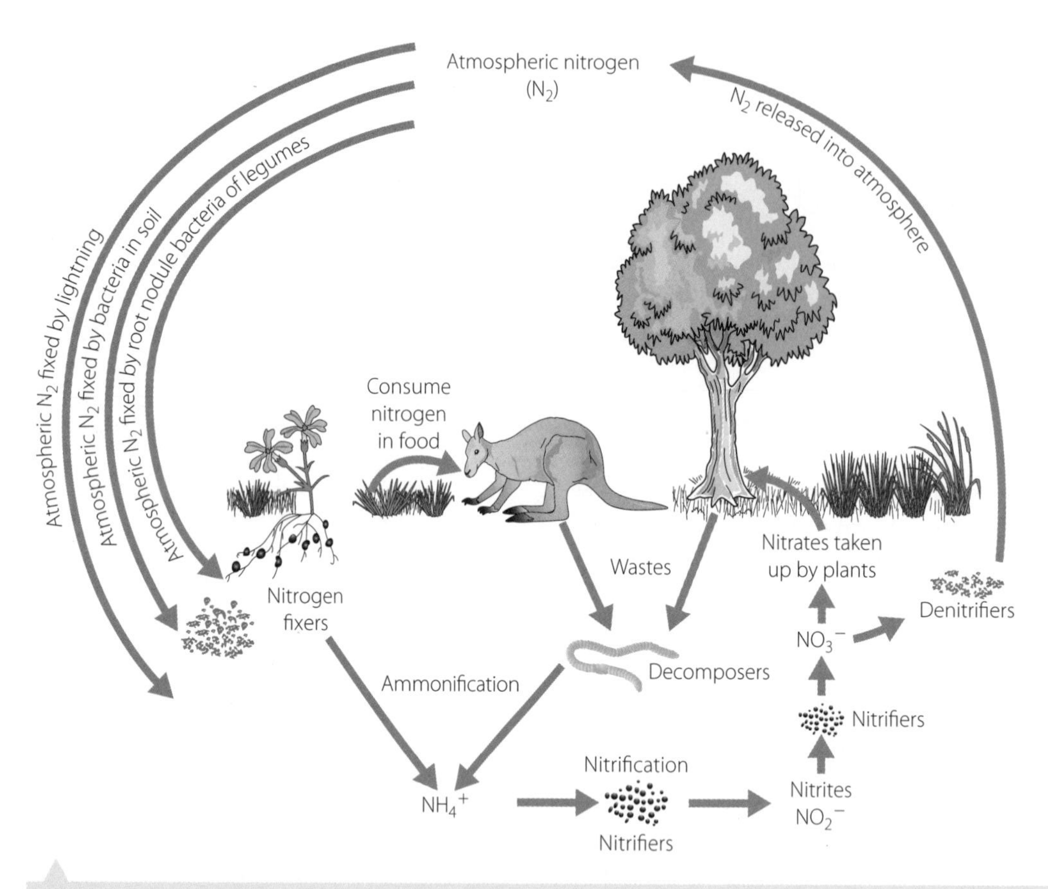

FIGURE 3.5.1 The balance between atmospheric nitrogen and ionic nitrogen is maintained by a set of key bacteria, including nitrogen fixers and denitrifiers.

QUESTIONS

1 Outline three major differences between the nitrogen cycle and the carbon cycle.

9780170411745

2 Excessive antibiotic use has resulted in copious quantities of antibiotics accumulating in the soil and water of our environment. Explain the implications of this fact for the future of the nitrogen cycle.

3.6 The water cycle

1 Describe how the water cycle can be imbalanced by an increase in global temperature.

2 Give three examples of negative environmental impacts of an imbalanced water cycle.

3 Explain whether an imbalanced water cycle could eventually rebalance.

3.7 Ecological niche

Ecosystems across the world are unique and diverse in their own ways. The organisms that inhabit a particular ecosystem are able to survive because of the particular set of biotic and abiotic factors present. The way in which species function within their environment, for example, the time they feed, what they feed on, where they live and when they reproduce, is known as an ecological niche. To place this concept into context, Eugene Odum (1913–2002), an American biologist at the University of Georgia, made the analogy that if the species habitat was its home address, then its ecological niche was its profession in that location. If two species attempt to occupy the same niche, that is, practice the same profession in the same location, one will eventually out-compete the other until only one remains.

QUESTIONS

1 Complete the table to define the ecological niche of the great white shark and the humpback whale.

	GREAT WHITE SHARK	HUMPBACK WHALE
Territory Where are they found? How deep do they range?		
Feeding behaviour What do they eat? How do they eat it? Is there anything they won't eat?		
General activity When do they sleep? When do they eat? Do they fight/defend territory?		
Social/reproductive behaviour How many are found together? How and when do they reproduce? How many offspring do they have?		

9780170411745

3.8 Data analysis: the competitive exclusion principle

The competitive exclusion principle postulates that no two species can occupy the same niche in the same environment for an extended period of time. Experiments can be used to test this principle and produce models of this interaction. Models can then be used to make predictions about biodiversity when changes occur within an ecosystem.

A team of scientists investigated Gause's competitive exclusion principle by conducting a field experiment. They studied two sessile (fixed or non-moving) species, A and B, introduced onto an intertidal rock face. Over a period of eighteen months, population density data was collected within the test quadrats. Table 3.8.1 is a summary of the data.

TABLE 3.8.1 Population density of two species, A and B

TIME (MONTHS)	POPULATION DENSITY OF SPECIES A	POPULATION DENSITY OF SPECIES B
0	5	5
1	12	8
2	17	14
3	36	24
4	55	60
5	50	67
6	43	62
7	45	104
8	50	100
9	38	104
10	25	115
11	30	119
12	16	125
13	18	133
14	21	139
15	14	158
16	8	141
17	0	158
18	0	164

QUESTIONS

1 Sketch a rough graph of the tabled data.

2 Which species had the competitive advantage in the first three months?

3 For what period of time can the two species coexist in the same area?

4 Which species had the overall competitive advantage over the eighteen months?

5 Outline what has occurred to species A throughout the eighteen-month period.

6 Predict the consequences of this scenario on the biodiversity of an ecosystem.

7 Assess the validity of Gause's competitive exclusion principle when applied to this experiment.

8 If a disease were to affect the reproductive success of species B at 10 months, predict what would happen to the relative abundance of species A and B.

9 An introduced species, species C, was introduced to the intertidal rock face. It competes for the same resources as species A and B, but has a shorter gestation period than both species. Predict the impact this would have on the relative abundance of all three species.

10 Design an experiment to measure the impact of species C on the ecosystem.

3.9 Keystone species

A FOREST FED ON FISH

Grizzly bears make the forests of British Columbia, Canada, their home. Bear populations living by the rivers have a distinct advantage over other bears in the region – they can catch and eat fish (Figure 3.9.1). In autumn, large numbers of salmon travel up the river from the ocean to their spawning areas to produce their young and are in plentiful supply for hungry bears. The bears can eat 40 kg of fish per day for several weeks, gaining considerable weight. This is essential for them to survive over the long winter hibernation. Bears are key to the success of the forest ecosystem and occupy a crucial link in the food web connecting the river and the forest. The fish carcasses are discarded along the river banks and forest floor, which provides a breeding ground for insects and nutrients for their larvae. Decomposers finish the job of breaking down the carcasses to recycle all the nutrients to the soil. Scientists have estimated that 80% of the bears' composition is derived from the ocean. How can this be when the only source of ocean food is the salmon they consumed for a few weeks?

Getty Images/Ian Mcallister

FIGURE 3.9.1 Grizzly bears feed insatiably on salmon for several weeks of the year before hibernation.

In spring, the bears emerge from hibernation very hungry and consume large amounts of the new grasses that have grown. Scientific tests on these grasses reveal high levels of a rare nitrogen isotope predominantly found in the ocean, nitrogen-15 (^{15}N). It was the fish fertiliser scattered on the forest floor by the bears in autumn that provided the nutrients for the grasses to flourish and the bears to eat in spring. This accounts for the seemingly large amount of ocean material in the
bears' bodies.

The larvae that were incubated in the fish carcasses have matured into insects and are food for birds, which then carry this ocean-fed material further into the forest through their droppings. Scientists have removed samples of wood from established trees and have measured the amounts of ^{15}N, relating it to the growth rings. This data has shown that the presence of ^{15}N, and therefore salmon fertiliser, is dating back several decades.

The bears' diet has sustained an entire ecosystem.

QUESTIONS

1 Draw a diagram of this food web and identify the primary, secondary and tertiary consumers.

2 What is the main feature that has allowed this ecosystem to succeed in a region that would have otherwise looked very different?

3 Explain the source of the large amount of ^{15}N in the bodies of the bears.

4 In this ecosystem, bears are at the top of the food chain and have large amounts of ^{15}N. What ecological process explains the large amount of this isotope in the bears' tissues?

5 Find out where salmon appear in the food chain and then draw a biomass pyramid to include bears.

6 Predict what would happen to this food chain if the bears all died from disease. You may construct a diagram to show this.

1 Energy and matter in ecosystems can be depicted in all of these, except:

A Food chains

B Cladograms

C Nutrient cycles

D Biomass pyramids

2 Energy is lost in each energy transfer through:

A Radiation and respiration

B Respiration and photosynthesis

C Conduction and convection

D Growth and development

3 Describe how an energy flow diagram is different to a food chain.

4 Calculate the energy available to the top-order predator in the following food chain.

Sun → Producers – 2000 kJ → 1st Consumers → 2nd Consumers → 3rd Consumers

5 Draw an energy flow diagram for an ecosystem that has four trophic levels, receives 10 MJ of light energy from the Sun and loses 7% to heat loss and 6% to decomposers at each trophic level.

6 Define ecological niche.

7 Given the extreme adaptability of humankind as a species, our fundamental and realised niches are the same. Defend or refute this statement.

8 The southern cassowary (*Casuarius casuarius*) is considered a keystone species in the northern Queensland rainforests. They eat fruits from more than 100 rainforest plant species, and are the main seed disperser for most of them. Predict what would happen to the animals and plants in the rainforest if the cassowary died out.

9780170411745

4 Population ecology

Summary

- Carrying capacity indicates the maximum population size of a species that can be supported in a given environment. The carrying capacity depends on the biotic and abiotic limiting factors at any given place and point in time.
- A population is increasing if the birth rate and/or immigration rate exceeds the death rate and/or emigration rate. A population decreases when the death and emigration rates exceed birth and immigration rates.
- Population growth rate = (birth rate + immigration rate) − (death rate + emigration rate).
- Populations can be measured directly, for example through satellite images, fish nets, aircraft, GPS tracking and drones.
- Sampling techniques provide estimates of population size. Samples of a group of individuals, selected from the total population in a given area or volume, are taken to represent the total population.
- The Lincoln Index is commonly used to sample mobile species. This technique involves capturing individual animals, marking them and then releasing them. After a time, the population is re-sampled and the number of marked animals caught gives an indication of population size.
- There are patterns in the way populations are distributed. In random distribution, individuals are spaced irregularly. In uniform (continuous) distribution, individuals are evenly spaced. In clumped (grouped) distribution, a number of individuals are grouped together.
- When biotic and abiotic resources are abundant, populations have the ability to expand rapidly. This results in exponential population growth resulting in a J-shaped growth curve when population size is plotted over time.
- For a given species in a particular habitat, there is a certain equilibrium population that the ecosystem can support. In the logistic population growth model, the rate of population increase approaches zero as the population size nears the carrying capacity. When the logistic model is graphed it produces an S-curve.

4.1 Carrying capacity of populations

There is a limit to the number of individuals that can occupy an environment. As a population's size increases, the demand for resources, such as food, water, shelter, and space, also increases. Eventually, there will not be enough resources for each individual. The carrying capacity is reached.

QUESTIONS

1 Define 'carrying capacity'.

2 The female kangaroo is able to manage the number of young that are born at any one time. The female may have up to three young at various stages of development. When resources are plentiful all the young are likely to survive. When resources are scarce the female is able to abandon any one of the three.

List the biotic and abiotic limiting factors that determine the carrying capacity of kangaroos.

TABLE 4.1.1

BIOTIC LIMITING FACTORS	ABIOTIC LIMITING FACTORS

3 List up to seven natural catastrophic events that could affect the carrying capacity of populations in their environment.

9780170411745

Refer to the following information for questions 4 to 7.

In this scenario kangaroos have gained access to an area that was previously fenced off. After one year, four kangaroos have moved into the area. The number of kangaroos was recorded for thirteen years.

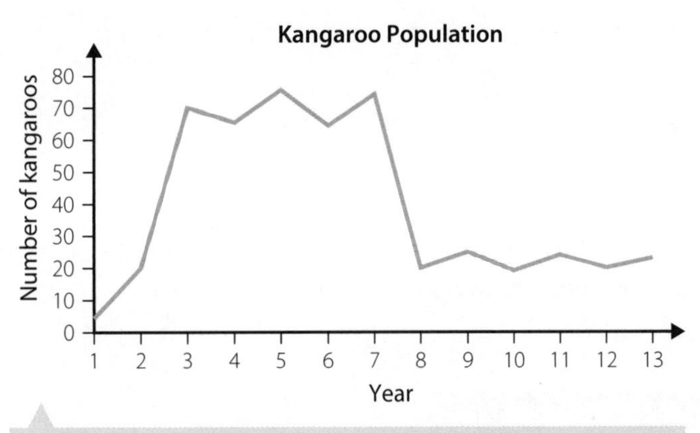

FIGURE 4.1.1 Data on the kangaroo population

4 Describe what is happening between years one and three.

____

____

5 Estimate the carrying capacity of the kangaroo population in this area up to year eight.What evidence did you base your answer on?

____

____

6 Suggest some factors that could have caused the decrease in kangaroo numbers at year eight.

____

____

____

7 Did the carrying capacity change in the last four years? Give reasons for your answer.

____

____

4.2 Population changes

Ecologists observe the rate of births, deaths, immigration and emigration of individuals to calculate the population growth rate. The growth rate can then be used to estimate the population size over time.

To find the population growth rate, ecologists use the formula:

$$\text{Population growth rate (PGR)} = (br + ir) - (dr + er)$$

where, *br* is the number of births per specified number of individuals, *ir* is the number of individuals that immigrate to the population per specified number of individuals, *dr* is the number of deaths per specified number of individuals, and *er* is the number of individuals that emigrate from the population per specified number.

QUESTIONS

1 In a population of 100 sparrows, 11 are born, 4 die, 5 immigrate and 7 emigrate every year.

a Calculate the population growth rate as a number of individuals and as a percentage. Show your working.

b At time zero (T_0), the population of sparrows is measured at 135. Using the population growth rate calculated in part a, estimate the population size over five years in the table below.

TABLE 4.2.1

TIME	POPULATION SIZE
T_0	135
T_1	
T_2	
T_3	
T_4	
T_5	

2 In a population of 1200 fish, 93 are born, 42 emigrate, 36 die and 22 immigrate every year.

a Calculate the population growth rate as a number of individuals and as a percentage. Show your working.

b At T_0, the population of fish is measured at 1564. Using the population growth rate calculated in part a, estimate the population size over three years in the table below.

TABLE 4.2.2

TIME	POPULATION SIZE
T_0	1564
T_1	
T_2	
T_3	

c In the third year, researchers noticed the fish population did not reach their estimation. Suggest a reason for this.

9780170411745

4.3 Measuring populations

Measuring the size and distribution of populations is important to ecologists. Data collected can assist in the management of populations.

QUESTIONS

1 Direct observation of individuals in a population is possible but sampling is often a better way of determining population size. List two disadvantages of direct measurement.

2 Describe two methods that can be used to record observations of populations in inaccessible regions.

3 The Lincoln Index is commonly used to sample mobile species. The formula for calculating the estimated abundance of animals using this technique is as follows:

Total population size = number captured × number recaptured ÷ number tagged in recapture

Describe how this technique works.

4 Ecologists in Cairns are concerned about a drop in the numbers of turtles. The number has been decreasing steadily over the last sixteen years. The following data from a turtle nesting site was used as evidence of decreasing numbers.

TABLE 4.3.1

COLLECTION TIME	NUMBER OF TURTLES
16 years ago Three weeks later	150 captured, tagged and released 170 recaptured (50 tagged, 120 not tagged)
Currently Three weeks later	50 captured, tagged and released 52 recaptured (17 tagged, 35 not tagged)

a Calculate the population size 16 years ago and currently. Show your working.

b Describe the changes in the turtle abundance in the nesting site in Cairns.

c List three factors that could lead to inaccurate estimates of population size using the Lincoln Index method.

5 Distribution of populations can occur in three typical ways: random, clumped or uniform. The way a population is distributed within a physical space helps scientists understand how the population operates.

Draw the three types of distribution in the boxes below.

CLUMPED DISTRIBUTION	UNIFORM DISTRIBUTION	RANDOM DISTRIBUTION

4.4 Population growth patterns

When biotic and abiotic resources are abundant, populations have the ability to expand rapidly. This unlimited growth doesn't normally occur indefinitely as there will be fewer resources as the population increases. Ultimately there is a limit to the number of individuals that can occupy a habitat.

QUESTIONS

1 Bacteria replicate through binary fission. A bacterium can replicate every 2 minutes.

a Fill in the data table below.

TABLE 4.4.1

TIME	NUMBER OF BACTERIA
0	2
2	
4	
6	
8	
10	
12	
14	
16	
18	
20	
22	
24	
26	
28	
30	
32	

9780170411745

b Graph the data in the grid below.

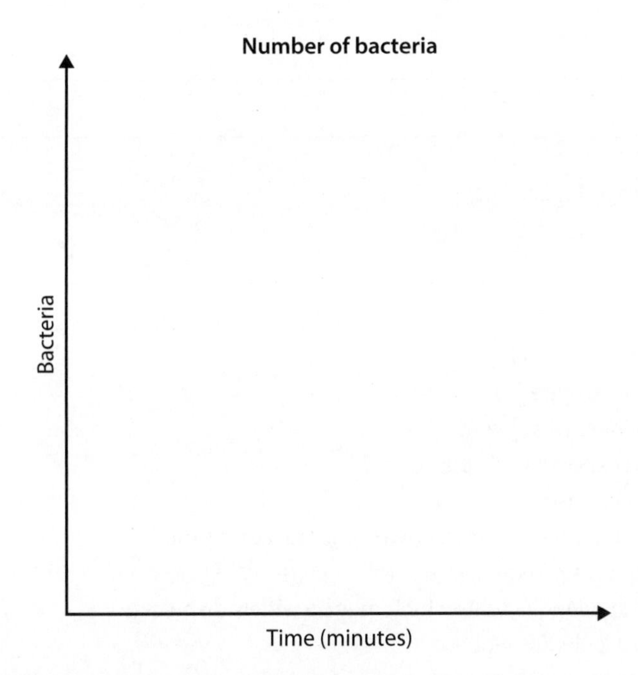

c Describe the type of population growth shown by the graph.

____

d Figure 4.4.1 shows the population growth when the bacteria are left for another fifteen minutes in their environment. Explain what is causing the change to population growth.

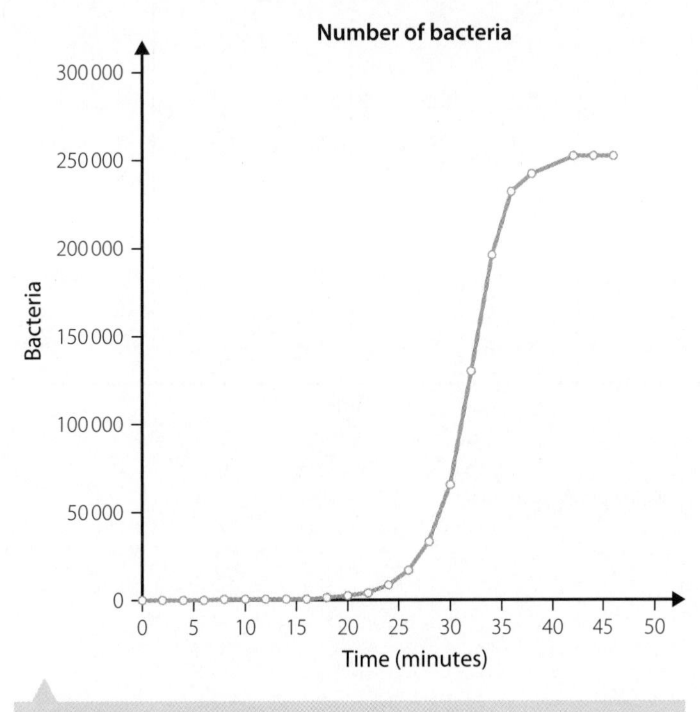

FIGURE 4.4.1 Population growth of bacteria

____

____

e Describe the type of population growth shown in this graph.

____

2 Describe the circumstances in which population growth rate would be similar to that in Table 4.4.1 and provide an example.

____

____

3 Add the following labels to Figure 4.4.2.

Environmental resistance starts – increasing death rate and/or decreasing birth rate
Population growing slowly phase
Birth rate exceeds death rate
Population growth increases slowly
Population growing exponentially phase
Constant population phase
Maximum growth rate under optimal environmental conditions
Birth rate and death around equal
Shortage of reproducing individuals which may be widely dispersed
Population growth deceleration phase

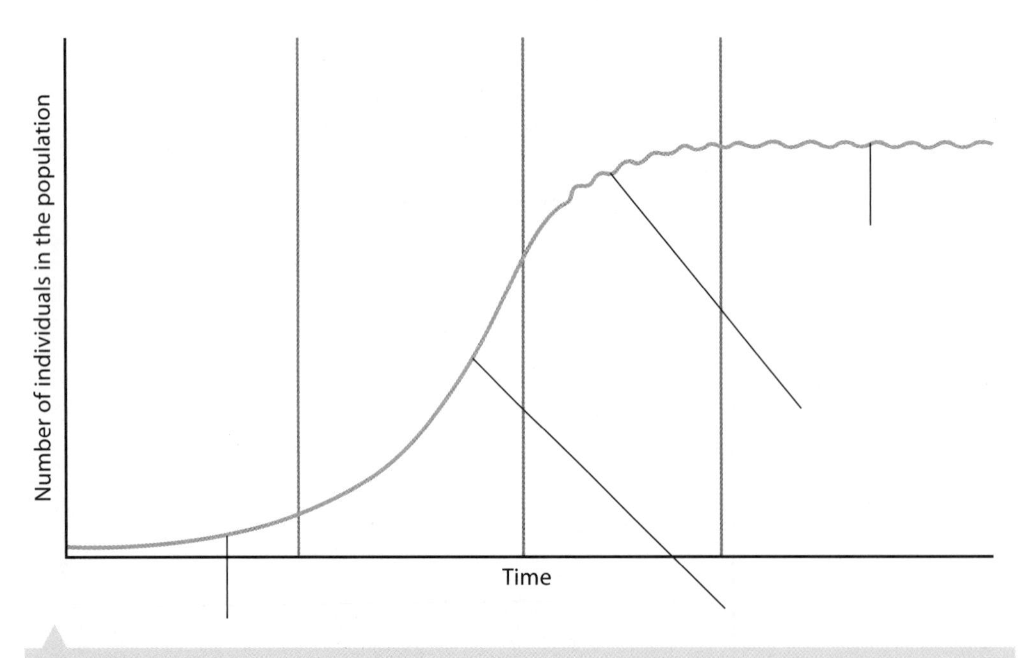

FIGURE 4.4.2 Population growth patterns

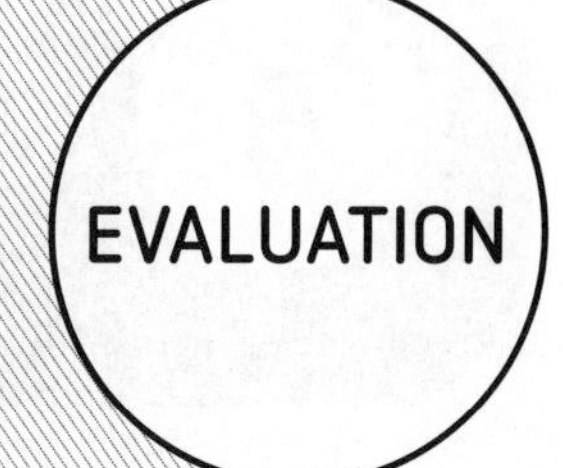

1 One area in a rainforest has been left undisturbed and an area close to it has been cleared for logging. Name the area in which there would be exponential growth of species. Explain your answer.

2 Explain why exponential growth is not sustainable in the long term.

3 The birth rate in a population of rainbow lorikeets is 0.07 per bird per year. The death rate in this population is 0.05 per bird per year.

a State whether the population is increasing or decreasing and calculate the number of individuals added to or lost from a population of 1000 rainbow lorikeets.

b Two other factors could add to or take away from the rainbow lorikeet population. Name these two factors.

4 Populations grow more slowly as they reach their carrying capacity. Describe two biotic limiting factors and two abiotic limiting factors that affect carrying capacity.

a Biotic limiting factors:

b Abiotic limiting factors:

5 Ecologists in Queensland estimated the population of red-bellied black snakes in an area by tagging 200 snakes and then releasing them back into the population.

a Write the name of the sampling technique and the formula used by the ecologists to estimate the population size.

b If 350 snakes were trapped one year later and 70 of those were tagged, calculate the total population. Show your working.

5 Changing ecosystems

LEARNING

Summary

- Communities change progressively over time, with one community being replaced by the next in serial replacement known as succession.
- Primary succession is the first stage of development of barren sites where no organisms inhabit the affected area.
- Pioneer plants are the first to colonise a barren area. They are fast-growing and typical of r-selected species. They are characterised by exponential growth making use of the resources and living space with little or no competition.
- Secondary succession occurs after events such as flooding, fire or logging. It is the re-establishment of a community where one previously existed.
- The end of succession is marked by a climax community made up of slow-growing, long-lived K-selected species.
- Movement of continents millions of years ago was an important stage in the physical isolation that allowed distinctive characteristics to develop in Australia's biota.
- Changes in Australia's ecosystems over time can be deduced from studying fossil records, soils, rocks and ice cores.
- The accumulation of ice layers in places such as Antarctica leaves an annual record of gases and dust in the atmosphere. These chemical traces can be retrieved in ice core samples and analysed to help build a more complete record of past climates.
- Models such as those demonstrating successional changes over space and time, are built using data gathered and the interpretation of that data.
- The First Australians survived and lived sustainably for tens of thousands of years.
- European colonisation dramatically increased the rate of change to the environment.
- Urbanisation, habitat destruction, land and soil degradation, salinity and monoculture practices are examples of the impact of humans on ecosystems today.

9780170411745

5.1 Ecological succession

Ecosystems are continually changing. Succession describes the way one species succeeds another in a gradual and progressive way.

QUESTIONS

1 Draw a flow diagram of primary succession after a volcanic eruption.

2 After flooding and heavy rains, a pond forms. Sediment run-off from surrounding land settles in the bottom of the pond. As sediment accumulates, the pond becomes shallower and plants start to spread towards the centre of the pond. With increased plant growth and litter the pond fills in to become a marsh. Paperback trees, which can tolerate waterlogged soils, grow and fill the area. The pond is converted into a paperbark scrub.

Name the type of succession described in this example and provide an explanation for your answer. Describe the climax community.

Type of succession: ______________________________

Explanation: ______________________________

____

Climax community: ______________________________

3 Figure 5.1.1 depicts the main features of a primary succession over time.

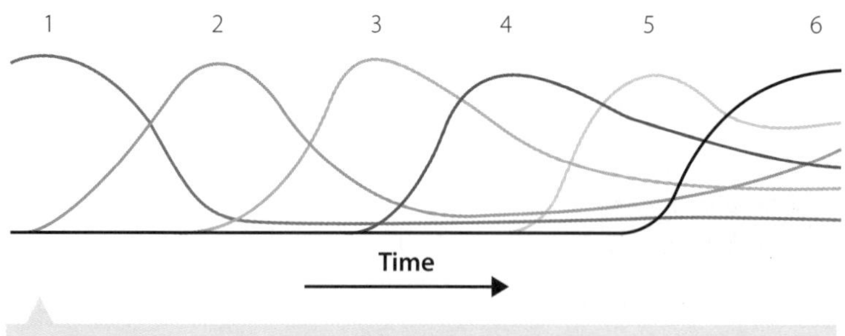

FIGURE 5.1.1 Forest succession over time in six stages

Create a key for the different features, labelled 1 – 6 in Figure 5.1.1, by matching them with the main features in the table below.

MAIN FEATURE IN FOREST SUCCESSION	NUMBER IN FIGURE
Grasses, perennials	
Climax forest	
Bare rock	
Fast-growing trees	
Woody pioneers	
Mosses, grasses	

4 Compare r-selected species and K-selected species by completing the table.

	r-SELECTED SPECIES	K-SELECTED SPECIES
Definition		
The stage they appear in ecological succession		
Why they are normally found in the stage you have identified		
Name of the population growth curve		

5 Distinguish between primary and secondary succession.

5.2 Fossil records

Scientists analyse evidence to reconstruct past ecosystems and map the changes in biotic and abiotic components that have occurred throughout Earth's history. The fossilised remains of living things, atmospheric changes trapped in ice cores and human activities such as rock paintings provide evidence of past ecosystems.

QUESTIONS

1 Around 40 000 years ago the West Kimberley region consisted of lush, open tropical forests. Approximately 10 000 years ago rainfall decreased and there was a cooling period, coinciding with the most recent glacial period. Scrub and open grasslands now dominate the area. With the end of the ice age, an Australian summer monsoon climate pattern emerged. Periodic droughts occurred when the monsoon failed.

Outline how Aboriginal rock paintings and fossils found in the area are used to provide evidence for analysis to justify this description of changes in the West Kimberley region. Describe how evidence in ice cores from Antarctica could further support this analysis.

2 Fossil remains of horse ancestors can be analysed to draw conclusions about the ecosystem they lived in. Shape and wear patterns on teeth can indicate the type of food and mode of eating of an animal. Animals living in dense forests with fruits and soft plants generally have small teeth with short crowns, they are smaller than animals in a more open habitat such as grasslands and they don't run fast. Animals in open grasslands and prairies have a diet of tough plants that require large, high ridged crowns on their teeth, and move fast. A single hoof allows for greater speed than a foot with digits.

Table 5.2.1 shows a modern-day horse and fossils from its ancestors. Using the information in the table, predict the ecosystem each species is likely to have lived in and how biotic and abiotic factors have changed over time. Use fossil records to justify your answers.

TABLE 5.2.1 The modern-day horse and its ancestors

TIME FRAME	DIAGRAM	DESCRIPTION
Equus The modern-day horse		• approximately 1.6 m high • a single visible digit • high crowned teeth extending into the sockets in the skull and jaw; span of cheek teeth 18 cm wide
Pliohippus 5.3–2.6 million years ago		• approximately 1 m high • single visible digit
Merychippus 23–5 million years ago		• increase in height of teeth crowns; span of cheek teeth 11 cm wide • three digits with side digits shorter than middle one
Mesohippus 34–24 million years ago		• approximately 60 cm high • four visible digits
Hyracotherium 55.8–33.9 million years ago		• 25–50 cm high • the teeth were low crowned with ridges; span of cheek teeth 4 cm wide • four visible digits on forefoot, three on hindfoot

3 Riversleigh in Queensland was once a lush rainforest. Today it is part of the driest vegetated continent on Earth. Explain how scientists could accurately predict and describe Riversleigh's ecosystem 25–15 million years ago.

____

____

____

____

5.3 Human impact on biodiversity

Extinctions result from reduced biodiversity. Studying a wide array of recorded extinctions and many species currently threatened with extinction, conservation biologists have identified human factors that seem to play an important role. Three factors identified are over-exploitation of resources, introduction of new species and disruption of ecological relationships.

9780170411745

QUESTIONS

1 Describe an example for each of the three factors identified.

2 Records from an Australian pastoralist in the mid-1880s indicate a holding of more than 11 000 sheep on a property of just under 5000 ha, a little more than two sheep per hectare. This is well above the normal carrying capacity for such a large herbivore. Sheep and other introduced large herbivores have hard hooves as opposed to the soft-footed structures of native animals, and so compact the soil when they graze. This creates opportunities for invasive, shallow-rooted, introduced plants at the expense of the deep-rooted native grasses. Farming practices that rely on large and heavy machinery for efficiency have added to the problem by compacting soil, as is evident when removal of topsoil by wind or water exposes the deep, hard ruts in paddocks. Tilling the soil is one way that farmers break up the soil for ease of planting. Traditionally, methods of tilling destroyed the superficial topsoil structure.

Explain how the agricultural practices described in the paragraph above increase the risk of soil degradation.

3 Complete the following table summarising the impact of humans on magnitude, duration and speed of ecosystem change.

TABLE 5.3.1

HUMAN IMPACT	MAGNITUDE	DURATION	SPEED
Urbanisation			
Habitat destruction			
Land and soil degradation			
Salinity			
Monoculture practices			

4 List three human activities, not included in Table 5.3.1, that are currently impacting on biodiversity.

1 Define:

a primary succession

b secondary succession

2 Describe an event that would precede:

a primary succession

b secondary succession

3 Describe the type of species found in a climax community.

4 Identify four features of pioneer species that suit them to their role.

5 Explain how analysis of ice core samples are used to contribute to the reconstruction of past environments.

6 A fossil is found in central Queensland. Describe three pieces of evidence that could be used to determine what the ecosystem was like at the time when the fossil formed.

9780170411745

7 Briefly describe three human practices that have reduced biodiversity in Australia.

---

8 A forested area has been cleared and redeveloped as prime agricultural land. Predict the impact of this on the ecosystem.

UNIT FOUR

HEREDITY AND CONTINUITY OF LIFE

- Topic 1: DNA, genes and continuity of life
- Topic 2: Continuity of life on Earth

9780170411745

6 DNA structure and replication

Summary

- James Watson and Francis Crick are credited with the discovery of the structure of DNA in 1953. Many groups of scientists have contributed to the study of the role of DNA.
- DNA is composed of four different types of nucleotides. Each nucleotide has a deoxyribose sugar, a phosphate group and one of four different types of nitrogen bases: adenine (A), thymine (T), guanine (G) and cytosine (C). The two strands of a DNA double helix are linked by hydrogen bonds between complementary bases: A links with T, G links with C.
- DNA replicates by a semi-conservative mechanism where one of the strands in the newly formed molecule is new and the other is the original strand. The enzymes DNA helicase and DNA polymerase facilitate DNA replication.
- When a eukaryotic cell prepares to divide, the DNA first replicates and then the chromatin condenses by coiling up, eventually becoming thick enough to be seen as a number of separate chromosomes. Replicated chromosomes contain two identical copies, called sister chromatids, joined by a centromere.
- In a non-dividing eukaryotic cell, DNA, coiled around histone proteins to form nucleosomes, is only visible as a diffuse grainy substance called chromatin.
- The DNA of both mitochondria and chloroplasts exists as a single, circular chromosome, similar to those of prokaryotes.
- Chromosomes in prokaryotic cells are generally circular and found in a region of the cell called the nucleoid. Small rings of DNA called plasmids may also be present in prokaryotic cells.

6.1 Chromosome structure

The genetic material in all living organisms is DNA. DNA, which is packaged into chromosomes, has the same basic structure in all organisms.

QUESTION

1 Fill in the missing words in this cloze activity.

In the nucleus of a ____________ non-dividing cell, DNA is only visible as a ____________ substance without ____________. Early microscopists named this seemingly diffuse, grainy substance ____________. ____________ is the cell's DNA together with all the ____________ associated with it. When the cell prepares to divide, the chromatin ____________ by ____________ up. It eventually becomes ____________ enough to be seen, when stained, as a number of separate structures called ____________. A chromosome is ____________ DNA molecule with its associated ____________.

For most of the time, chromosomes are ____________. An ____________ chromosome is a ____________, long DNA double ____________ molecule coiled around ____________ proteins. By contrast, ____________ chromosomes, which have undergone DNA ____________ in preparation for cell ____________, contain two ____________ copies, called ____________ chromatids. Sister chromatids are joined by a ____________, making them appear as an 'X' ____________.

6.2 Packaging of DNA

There is around two metres of DNA in every cell of a eukaryotic organism. Even prokaryotes contain relatively long lengths of DNA. Given the microscopic size of cells, this DNA must be packaged in some way.

QUESTIONS

1 Compare and contrast the DNA in eukaryotes with the DNA in prokaryotes.

2 Explain the difference between chromatin and a chromosome.

3 Explain why most images and drawings of chromosomes show them in the duplicated form.

---

4 Relate the concept of a chromosome to the concept of a gene.

---

5 Organise the following descriptions of the levels of organisation in a chromosome, from largest to smallest, by placing the numbers 1–5 in front of each statement.

______ The loops of coiled DNA are wound around a core of eight histone proteins to produce a nucleosome.

______ Interacting proteins package loops of coiled DNA into a 'supercoil', to produce chromatin, which is organised as a cylindrical fibre.

______ The double helix of DNA is held together by hydrogen bonds between nitrogenous bases.

______ A tightly coiled and condensed human chromosome is only visible when stained during cell division, after DNA replication.

______ A nucleosome consists of a section of DNA molecule looped twice around a core of histones.

6.3 Case study: what is a gene?

In 1928, a British army medical officer, Frederick Griffith, was trying to develop a vaccine against the strain of bacteria that causes pneumonia, *Streptococcus pneumoniae*. At that time, scientists understood the basic patterns of inheritance, but the physical nature of the gene was unknown. For many years, genetic information was thought to be contained in cell proteins.

In his experiment, Griffith isolated two different strains of bacteria: the harmless R strain without a capsule and the disease-causing encapsulated S strain that had a smooth surface. This polysaccharide capsule prevented phagocytosis by the host's innate immune cells. He injected laboratory mice with both strains and found that those injected with the R strain remained healthy and that the bacterial cells in their blood died, while those mice injected with the S strain died and their blood contained living bacterial cells.

When S strain bacterial cells were killed by exposure to high temperature and injected into mice, the mice remained healthy with no living bacterial cells found in their blood. However, when heat-killed S strain bacterial cells were combined with living R strain cells and injected into mice, death occurred and both live S and R strain cells were found in their blood.

How can these observations be explained? The simplest idea is that high temperatures did not affect the hereditary material in the S strain bacterial cells and so this material was transferred to the living R cells, transforming them so they became deadly. This showed that the hereditary material was not protein, which is easily denatured by heat.

Some years later, Oswald Avery took Griffith's experiments one step further. S strain bacteria were treated with protease enzymes, which removed the proteins from the cells before being placed with R strain bacteria. The R strain bacteria transformed, showing that proteins did not carry the pathogenic genes. When the S strain bacteria were treated with a nuclease enzyme to remove the DNA, the R strain bacteria no longer transformed. This confirmed that DNA was the hereditary material in cells.

QUESTIONS

1 Recall the meaning of pathogenic.

2 Describe the structure and function of a gene.

3 Describe two ways in which a genome differs from a gene.

4 Explain how Frederick Griffith's observations with *S. pneumoniae* contributed to Oswald Avery's conclusion that hereditary material was made up of nucleic acids.

5 When they described the structure of DNA, Watson and Crick noted that the specific base pairing they proposed, immediately suggested a possible copying mechanism for the genetic material. Explain what they meant by that statement.

9780170411745

6.4 DNA and its replication

1 Each of the following statements is incorrect. Rewrite it as a correct statement.

a One strand of the DNA helix ladder is maternal and the other strand is paternal.

____

b Different organisms have different types of DNA because they are very different from each other.

____

c Each chromosome is made of more than one DNA molecule.

____

d The different cell types (skin, muscle, cartilage, etc.) found in a given individual's body contain different DNA.

____

e In sexually reproducing organisms, half of the organism's body cells contain DNA from the mother and half contain DNA from the father.

____

____

Use Figure 6.4.1 to answer the following questions.

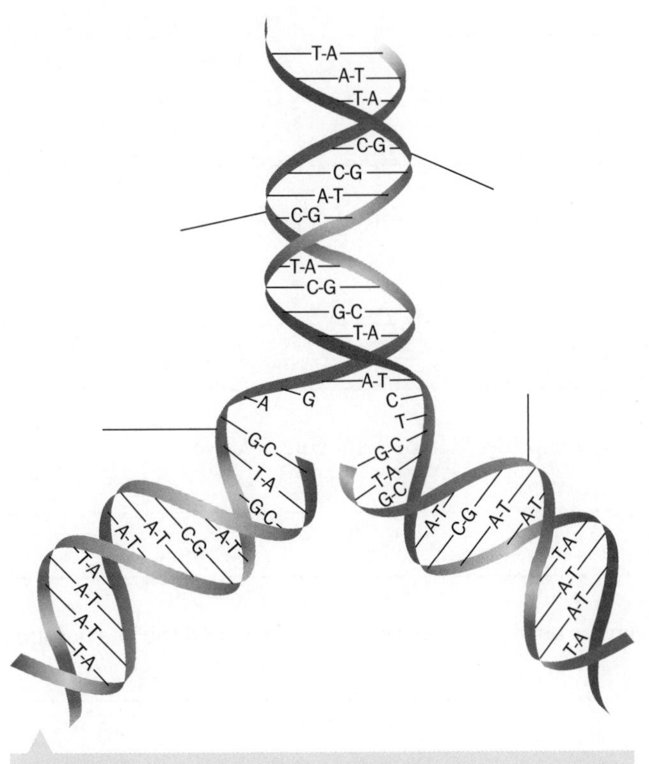

FIGURE 6.4.1 DNA preparing for cell division

2 Explain why DNA is referred to as the 'double helix'.

3 State what the letters A, C, T and G represent in the diagram.

4 State the chemical components of a nucleotide.

5 Describe how the two strands of DNA are held together.

6 Label the parts of the diagram indicated with pointers.

7 Describe what is happening in the diagram. What is the name of this process?

8 Where do the new pieces of DNA (replicate strands) that join to the original pieces (parental strand) of the DNA strand come from?

9 Explain why DNA replication is considered to be 'semi-conservative'.

10 Name the junction between the unwound single strands of DNA and the intact double helix.

9780170411745

11 Describe DNA replication, including the roles of the two enzymes that are involved.

1 Which of the following statements is incorrect:

A RNA is composed of a single chain of nucleotides.

B In DNA the base thymine is replaced by uracil.

C RNA has important functions in protein synthesis.

D DNA is a nucleic acid related to RNA.

2 Choose the correct statement:

A All living things inherit half their DNA from each of their parents.

B DNA is found in the chloroplasts and mitochondria of eukaryotes and prokaryotes.

C DNA contains instructions for the growth and functioning of all organisms.

D In prokaryotic organisms, DNA exists as a single stranded, circular chromosome.

3 If 30% of the bases of a molecule of DNA are cytosine, it will also contain

A 15% guanine

B 30% thymine

C 40% uracil

D 20% adenine

4 Describe the semi-conservative replication of DNA.

____

____

____

5 Distinguish between the roles of DNA helicase and DNA polymerase.

____

____

____

____

9780170411745

6 DNA is the genetic material of all cells, and is also responsible for controlling cellular activities. Explain how the structure of DNA enables it to effectively transmit information from one generation to the next and describe how this information is copied prior to cell division.

7 Cellular replication and variation

Summary

- In eukaryotes, somatic or body cells have pairs of homologous chromosomes; one chromosome of a pair comes from the male parent and the other chromosome of a pair comes from the female parent.
- Homologous chromosomes have the same genes at the same position (locus) but these genes may have alternative forms of the gene called alleles. Sex chromosomes that determine an individual's sex are generally matched in one sex (e.g. XX) and unmatched in the other sex (e.g. XY).
- When matched and ordered, eukaryotic chromosomes are displayed in a karyotype and different chromosome sizes, centromere positions and banding patterns can be observed.
- A diploid number (2n) of chromosomes is found in somatic cells; a haploid number (n) of one of each of the pairs of homologous chromosomes is found in gametes.
- Meiosis is a form of eukaryotic cell division that is involved in sexual reproduction. In meiosis, nuclear division is followed by cytoplasmic division called cytokinesis. This results in the formation of four daughter cells, each containing half the number of chromosomes of the original nucleus.
- At fertilisation, two sex cells, also called gametes, usually from different individuals, combine to restore the original chromosome number, in a new cell, called a zygote, with genetic material from two different parent cells.

 9780170411745

7.1 Karyotypes, meiosis and hybridisation

Meiosis occurs in all sexually reproducing organisms. Its central role is the production of gametes.

QUESTIONS

1 Fill in the missing words in this cloze activity.

Microscopic examination of a stained __________ cell in the process of nuclear __________ reveals a jumbled cluster of __________ that exist in pairs, called __________ pairs. The exception is the __________ chromosomes which in one sex, usually the male, are not __________. __________ chromosome of each __________ pair is inherited from each __________, with members of each homologous pair sharing characteristic __________ patterns.

A __________ is the standard presentation form used to __________ and analyse chromosomes. Photographic __________ of __________, during cell __________, are arranged into __________ and ordered __________ to create a karyotype. The chromosomes are ordered by __________, from largest to smallest.

2 Describe how gametes differ from typical body cells.

____

____

____

3 Explain why meiosis is appropriate for gamete formation.

____

____

____

4 Describe two ways in which karyotypes of sperm cells would differ from those of the male they came from.

____

____

5 Complete the following table of chromosome numbers in various species.

SPECIES	2n	NUMBER OF HOMOLOGOUS CHROMOSOME PAIRS	n
Human	46		23
Fruit fly	8		
House fly			6
Chimpanzee			24
Camel	70		
Chicken		39	
Goat			30
Petunia		7	
Rice	24		

6 When animals of different species are kept together in captivity they sometimes mate and produce offspring. A donkey is known to have a diploid number of 62 and a zebra has a diploid number of 44.

a Name cells in the donkey that would be expected to contain 31 chromosomes.

b Name cells in the zebra that would be expected to contain 44 chromosomes.

c Estimate how many chromosomes are expected to be in the somatic cells of the donkey.

d If a 'zonkey' (a hybrid formed by the fertilisation of a female donkey egg with a zebra sperm) is produced, predict the 2n number.

e Describe how a karyotype is made and identify how chromosomes are ordered.

f Describe how a zonkey karyotype would differ from the karyotype of a zebra.

g Suggest problems that might occur when the zonkey produces gametes.

h Explain why most hybrid animals are infertile.

9780170411745

7.2 Aphids: an unusual reproductive strategy

Aphids, which are sap-sucking garden pests, have an unusual reproductive strategy. The overwintering eggs that hatch in spring have XX sex chromosomes, resulting in only females. Once fully grown, each female produces thousands of descendants. Eggs are produced, without meiosis and fertilization, and the offspring resemble their mothers. This kind of reproduction is called parthenogenesis. The process is repeated throughout the summer, producing multiple generations of females.

In autumn, a change in temperature, causes females to parthenogenetically produce eggs that give rise to sexual females and males. The males are genetically identical to their mothers, except that one X chromosome is eliminated from the egg cell before it develops into a male, which is XO. Sexual females and males produce sex cells in the usual way, except that only sperm containing an X chromosome survive. Any sperm without an X chromosome degenerate. The fertilised eggs hatch in spring and the lifecycle begins again.

QUESTIONS

1 State whether the eggs produced in autumn that overwinter are haploid or diploid.

2 Name the type of cell division that gives rise to eggs and sperm and state the number and type of sex chromosomes each would carry.

3 Explain why overwintering eggs only hatch into females.

4 Compare and contrast sex determination in humans and aphids.

5 Explain why it is beneficial that half the sperm degenerate before maturing.

7.3 Oogenesis and spermatogenesis

1 Fill in the missing words in this cloze activity.

The process of oogenesis begins in the __________ of __________ during embryonic development, before a __________ is born. There, __________ oocytes begin __________, but remain in prophase I until the female matures __________. After that time, a primary __________ completes meiosis I each month to form a __________ oocyte and a structure called a __________ body. __________ is unequal with almost all of cytoplasm going into the __________ oocyte. The polar body __________. The second meiotic division, which produces a __________ ovum (__________) and a __________ polar body, occurs only if a __________ fertilises the __________.

The production of sperm in __________ is called __________. Stem cells in the __________ undergo mitotic division, each time producing a new __________ cell that continues to divide, and a __________ primary __________. The latter divides in __________ I to form two __________ spermatocytes which in turn divide in __________ II to form four __________, which are __________ and develop into __________ sperm cells. This process occurs __________ a male's lifetime and is capable of producing, in humans, at least 3 __________ sperm per day.

In the process of __________, __________ sex cells fuse to produce a diploid __________. Human __________ produced by meiosis each contain n = 23 chromosomes. __________ restores the chromosome number to 2n = 46. Different species have __________ numbers of __________.

7.4 Case study: a unique system of sex determination

A unique system of sex determination is found in alligators and crocodiles. Here the sex of the offspring is determined entirely by environmental factors. There are no differences between males and females in terms of their chromosomes. Instead, there is a critical period in the early development of the fertilised egg during which the surrounding temperature determines the sex of the individual. In some species, high temperatures lead to males; in others, high temperatures result in females.

QUESTIONS

1 Compare and contrast how sex is determined in humans and in crocodiles.

2 Describe crossing over in humans.

3 What is the benefit of crossing over?

4 Describe independent assortment in humans.

5 What is the benefit of independent assortment?

6 Predict one way in which the system of sex determination, described here for alligators and crocodiles, could increase genetic variation in their populations in an area.

7 Suggest how the karyotype from a crocodile would differ from that of a human.

EVALUATION

Questions 1-3 use the following information.

A common species of moss has a diploid number of 10. The moss life-cycle starts when a haploid spore germinates to produce a familiar moss plant, which is composed entirely of haploid cells. Sex organs develop from the tips of this plant, with the female organ producing an egg and the male organ releasing sperm. Fertilisation produces a zygote, and some months later, spore-producing cells undergo meiosis to form haploid spores, and the cycle starts again.

1 Which of the following statements is correct?

A Meiotic division in spore-producing cells produces spores with 10 chromosomes.

B Cells produced in the sex organs of a moss plant have 5 chromosomes.

C All cells in the moss plant contain 20 chromosomes.

D Before meiotic division, each cell in the moss has 10 pairs of chromosomes.

2 Sperm are formed from:

A meiosis in the male sex organ.

B spore-producing cells.

C meiosis in the zygote.

D moss plant cells.

3 Spore-producing cells are:

A haploid like moss plant cells.

B diploid like the zygote.

C haploid like eggs.

D diploid like the moss plant.

4 State two ways in which sexual and asexual reproduction are different.

____

____

____

9780170411745

5 Describe three differences between oogenesis and spermatogenesis.

6 Crossing over is an important process in both plants and animals.

a State where and when crossing over occurs in humans.

b State whether crossing over increases or decreases variation in offspring.

c Explain how crossing over alters variation.

8 Gene expression

Summary

- The genome includes all cellular DNA and is unique to each individual organism.
- Genes are DNA sequences (coding DNA) that instruct polypeptide production.
- The specialised structure and function of a cell is determined by which proteins it produces.
- Proteins are comprised of two or more polypeptides.
- All cells in an organism have the same genome, but selective switching on (*expression*) of particular genes allows each specialised cell type to synthesise only the proteins necessary for that cell to function.
- Protein synthesis occurs through a series of steps:
  1. *Transcription*: a singular DNA template strand acts as a pattern or code for the production of messenger RNA (mRNA). Complementary nucleotide base pairings occur between cytosine (C) and guanine (G), and adenine (A) and thymine (T). On the RNA stand, thymine is replaced with uracil (U).
  2. After the non-coding *intron* segments are removed, mRNA moves out of nucleus and into the cytoplasm where ribosomes attach to mRNA and direct translation.
  3. *Translation*: transfer RNA (tRNA) molecules, each with particular three base *anticodons*, are attracted to complementary mRNA *codons*, thus delivering attached amino acids to a growing polypeptide chain. Different combinations of the 20 amino acids form different polypeptides.
- The vast majority of the genome contains non-coding DNA. Not all functions are known, but some are transcribed into tRNA and ribosomal RNA (rRNA), necessary for protein synthesis; some non-coding DNA (introns, for example) *regulates* genes, that is, causes certain genes to be expressed at certain times. Centromeres, which hold sister chromatids together in cell division, and telomeres, which protect chromosome ends, are also examples of non-coding DNA.
- There are many factors that impact gene regulation, through activation or deactivation.
- At transcription, regulation occurs as a result of chemical groups bonding to DNA or to histone proteins of chromosomes, and by regulatory proteins from other genes bonding to DNA.
  For example, acetyl chemical groups activate genes, while methyl groups deactivate genes.
  X chromosome-inactivation occurs in tortoiseshell cats by the attachment of methyl groups.
- At translation, regulation occurs when proteins bond to mRNA to block translation by ribosomes.
  Environmental conditions also affect gene expression.
- For example, low temperature darkens rabbit fur colour for heat absorption while heavy metals cause cellular malfunction.
- *Epigenetic* chemical regulation is passed from parent to offspring. For example, mouse offspring agouti gene is switched off due to the pregnant mother's high methyl diet.
- Embryonic development and sex determination are controlled by regulatory proteins.

9780170411745

8.1 Genome and genes

1 Complete the simple flow chart to illustrate the hierarchical relationship between *genome, genes, DNA strands* and *nucleotide bases.*

____________ → ____________ → ____________ → ____________

are part of   are part of   are part of

8.2 Protein synthesis

ACTIVITY

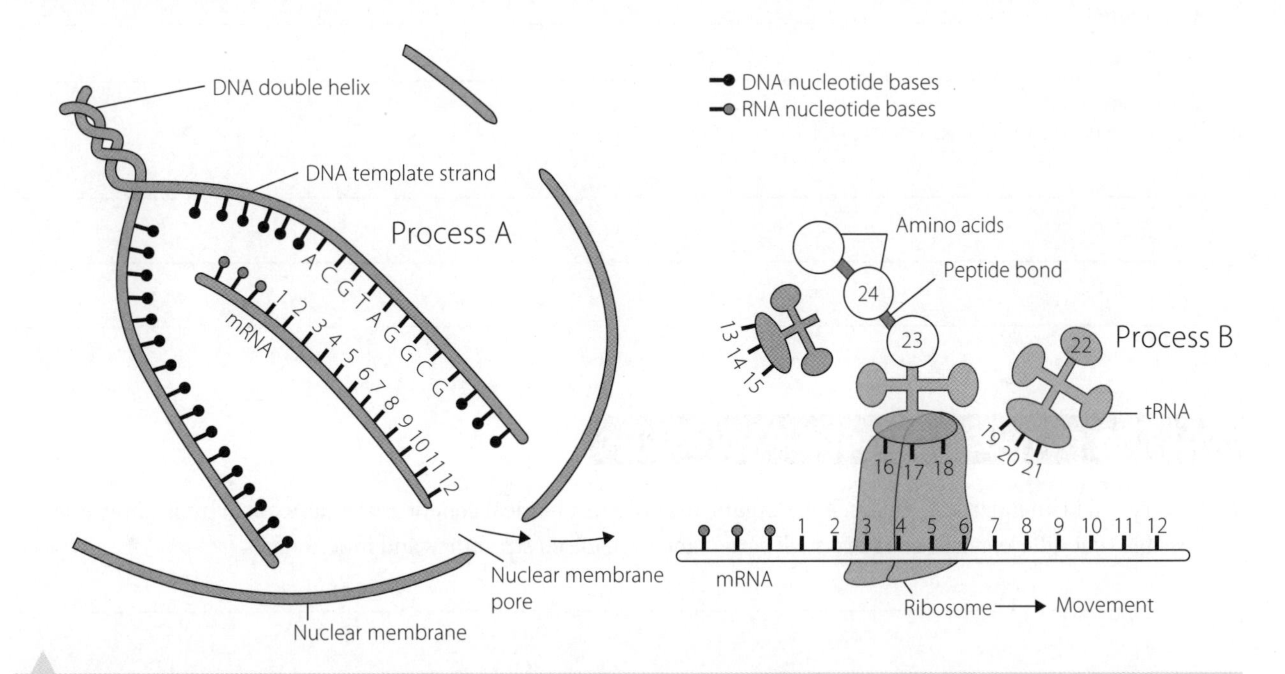

FIGURE 8.2.1 Schematic diagram representing protein synthesis for a short segment of DNA.

Answer the following questions, referring to Figure 8.2.1.

1 For 1–9, write the letter A, C, G or U to identify each of the nucleotide bases on the mRNA strand.

1______ 2______ 3______ 4______ 5______ 6______ 7______ 8______ 9______

2 For 13–21, identify each of the nucleotide bases forming anticodons on tRNA molecules if 13, 14, 15 have just disengaged from 1, 2, 3; 16, 17, 18 are about to engage with 4, 5, 6 and 19, 20, 21 will engage with 7, 8, 9.

13 ________ 14 ________ 15 ________ 16 ________ 17 ________ 18 ________ 19 ________ 20 ________ 21 ________

Refer to the genetic code, either in your textbook or from an online source, to answer the following.

3 For 10–12, identify what the nucleotide bases may be if together the three are a STOP codon.

10 ________ 11 ________ 12 ________

4 For 22–24, refer to the genetic code to determine the amino acids being added to the polypeptide chain.

22 ______________ 23 ______________ 24 ______________

5 For A and B, name the processes occurring and the location within the cell.

A: process ______________________, location ______________________

B: process ______________________, location ______________________

8.3 Coding and non-coding DNA

1 Define:

a coding DNA ____

b non-coding DNA ____

2 List three functions of non-coding DNA.

____

____

____

8.4 The purpose of gene expression

1 If every cell in an individual multicellular organism contains identical genetic material, explain briefly how it is possible that different cells can have such varied and specialised structures and functions.

____

____

____

____

____

2 Define:

a gene expression ______

b gene regulation ______

3 Write T (true) or F (false) beside the following statements:

a It is thought that the default state of gene expression is ON; that is, all genes are naturally 'switched on' unless a regulating factor turns them off. ______

b There are some genes that are permanently switched ON. ______

8.5 The factors regulating the phenotypic expression of genes

Phenotypic expression of genes is regulated by several different factors internal to cells, which may operate during *transcription* and *translation* stages of protein synthesis.

Factors include *non-protein chemical groups*, that act on histone proteins or DNA in chromosomes, to switch genes on or off; and *regulatory proteins* produced by genes at other parts of the genome.

Also, *environmental factors* external to organisms, can affect gene expression. *Environmental contaminants, temperature, diet, stress levels*, and for plants, available light and day length are all thought to influence gene expression. Some chemical environmental regulation can be passed on to the next generation, as *epigenetic regulation.*

QUESTIONS

1 Draw a diagram to illustrate how DNA is 'packaged' as chromatin in chromosomes. Include labelling for the DNA strand, nucleosomes and histone proteins components.

Explain why genes contained within chromosomes are 'switched off' or unable to be expressed. Refer to your diagram.

______

______

2 Create a mind map illustrating the connections between the terms in italics in this section's opening paragraph. Add any extra relevant details you can think of.

3 Add the following listed gene regulation examples as appropriate connections in your mind map.

a Himalayan rabbit fur colour change

b Agouti gene deactivation in mice

c Tortoiseshell cats' X chromosome-inactivation

8.6 Transcription factors regulate morphology and cell differentiation

Correctly complete the paragraphs following by choosing the correct word from the list.

embryonic	zygote	activate	expression
proteins	structures	sex	male
differentiation	morphology	fertilisation	regulatory

The particular body shape and form that an adult organism develops is called its ______________. Homeobox genes are a small number of genes that control ______________ development, including cell ______________. The expression of these genes is accomplished due to proteins produced in the egg before ______________. The concentrations of various of these ______________ are different at particular positions in the egg, and these concentrations are maintained in the ______________ after fertilisation, and in the early cell divisions of the embryo. These differing concentrations of maternal-effect proteins then act to either activate or repress ______________ of homeobox genes at different positions in the dividing embryo to stimulate development and growth of different ______________ at these specific positions, by in turn activating expression of many other genes in the genome.

Embryo gender is determined by the ______________ chromosomes present in every cell. In mammals, if a Y chromosome is present, the embryo develops into a ______________, due to the expression of the SRY gene on the Y chromosome that produces a ______________ protein that, as in the case of homeobox genes, binds to DNA to ______________ the expression of many genes to produce testes, and thus other male sex characteristics.

1 Which of the following statements is correct?

A A gene is part of a genome.

B A genome is part of a gene.

C A gene and a genome are different names for the same structure.

D Genes and genomes are both components of DNA.

2 Which of the following statements is correct?

A Translation occurs before transcription.

B Translation occurs in the cytoplasm.

C Transcription occurs in the cytoplasm.

D Transcription and translation both occur in the nucleus.

3 If the anticodon of a tRNA molecule is known, explain how to determine:

a the amino acid that would attach to the tRNA.

b the original codon on the DNA template strand that would code for this amino acid.

4 Explain why there are several codons for some amino acids but only one for others.

5 Distinguish between:

a coding and non-coding DNA

b gene expression and gene regulation

c mRNA and rRNA

____

____

6 Briefly describe the process by which regulatory proteins prevent gene expression at:

a transcription

____

____

b translation

____

____

7 Briefly describe an example of an environmental factor affecting gene expression:

____

____

____

____

____

9780170411745

9 Mutations

Summary

- All individuals in a species have the same chromosome set, yet show variation due to each individual's unique set of alleles (gene alternatives) inherited from parents. Individuals in a species have different phenotypes due to their unique genotypes/alleles.
- New alleles arise among individuals of a species due to changes in DNA: mutations.
- Mutations may occur spontaneously during DNA replication and during cell divisions.
- Environmental mutagens, including a variety of chemicals and various types of radiation, increase mutation rate.
- Some DNA replication errors cause mutations that affect just one nucleotide unit. This includes point mutations, and when one nucleotide is wrongly substituted. The effect of these mutations varies depending on which nucleotide is affected and whether it changes the single amino acid coded for.
- Other DNA replication errors cause mutations affecting whole sequences of nucleotide units. These errors include frameshift mutations: the mistaken insertion of one (or more) extra nucleotides, or deletion. The result is that all codons from that point onwards are different to those of unmutated DNA. Multiple wrong amino acids have a drastic effect on the polypeptides synthesised.
- Cell division errors cause mutations affecting whole chromosomes or large chromosome segments. These mutations have a drastic effect due to the huge number of genes involved. It results in aneuploidy, when the chromosome number is different to the usual diploid number (2n) in body cells or haploid number (n) in gametes. Both non-sex and sex chromosomes can be affected. This is caused by non-disjunction/non-separation of homologous chromosomes or sister chromatids, and produces trisomy (3 chromosome copies) or monosomy (1 copy) instead of 2 of each in body cells.
- Mutation inheritance can occur in which the mutation affects only daughter cells of the original mutated cell. Inheritance of mutations can only occur from gametes produced by meiosis, where the mutation is copied into every embryonic cell. Often, this kind of mutation results in spontaneous abortion of embryo/foetus. When a live birth occurs, there are often congenital disorders.
- Most inherited mutations are disadvantageous. Premature death before reproductive age usually prevents ongoing inheritance, but it is possible to be inherited if the individual's second allele is unmutated and can produce functional protein. Rarely, gene mutation produces an allele that enhances offspring or species survival, due to an improved synthesised protein increasing individual's ability to compete.
- Environmental factors combine with the inherited genotype to determine phenotype.

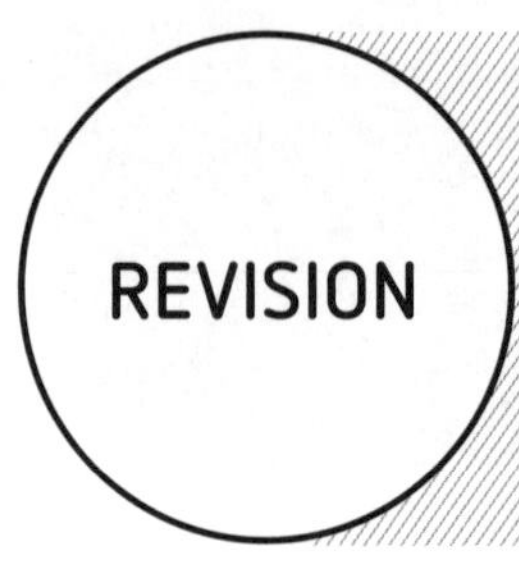

9.1 Identifying errors in genes and chromosomes

Mutations can be caused by errors in the DNA replication process that occurs between cell divisions. Firstly, recall the process of DNA replication without error.

QUESTIONS

Following is a *review* of the complementary base-pairing that occurs during DNA replication and in protein synthesis processes.

The base sequence for part of a DNA strand that will act as template for the replication of a new DNA strand is illustrated:

C-A-A-A-T-G-A-C-C-G-G-T-T-C-A-T-C-C

1 Write the sequence of the correctly replicated DNA strand.

____

____

2 From your answer to question 1, write the sequence of the resulting correctly translated mRNA strand, using the replicated DNA strand you wrote in 1 as a template.

____

____

3 By referring to the genetic code, either in your textbook or from an online source, write the sequence of the amino acids that would be translated from the mRNA strand in 2.

____

____

Following is a series of mutated DNA sequences that have been wrongly replicated from the DNA template sequence C-A-A-A-T-G-A-C-C-G-G-T-T-C-A-T-C-C.

QUESTIONS

For each mutated sequence:

- **a** select one or more of the following terms to describe the mutation type: point, substitution, phase-shift, insertion, deletion.
- **b** assume that the mutated DNA sequence acts as template for mRNA transcription, and write down the resulting mRNA base sequence.
- **c** write down the amino acid sequence resulting from translation (and based on b).
- **d** comment on the likely effect of the mutation on the resulting synthesised polypeptide, by referring to the *correct* amino acid sequence in Activity 1, part 3, that resulted from the unmutated DNA strand being replicated. Justify your answer.

1 G-T-T-A-C-T-G-G-C-C-A-A-G-T-A-G-G

a

b

c

d

2 G-T-T-T-A-C-T-G-G-C-C-A-A-G-T-A-G-C

a

b

c

d

3 G-T-T-T-A-A-T-G-G-C-C-A-A-G-T-A-G-G

a

b

c

d

4 G-T-T-T-A-C-T-G-G-C-C-A-A-G-T-A-A-G-G

a ______________________________

b ______________________________

c ______________________________

d ______________________________

____

____

____

____

9.2 Aneuploidy and karyotypes

Mutations can also be caused by errors in the cell division processes of mitosis and meiosis, when non-disjunction of homologous chromosome pairs (or large segments of the chromosomes at least) occurs at metaphase I, or non-disjunction of chromatids (or large segments of them) occurs at metaphase II. The homologous chromosomes or sister chromatids do not separate as they normally would, resulting in chromosome numbers other than the normal diploid (2n) number in somatic/body cells, or haploid (n) number in gametes. This condition is called aneuploidy.

QUESTIONS

Figure 9.2.1 illustrates a hypothetical cell with two pairs of chromosomes in prophase I of meiosis.

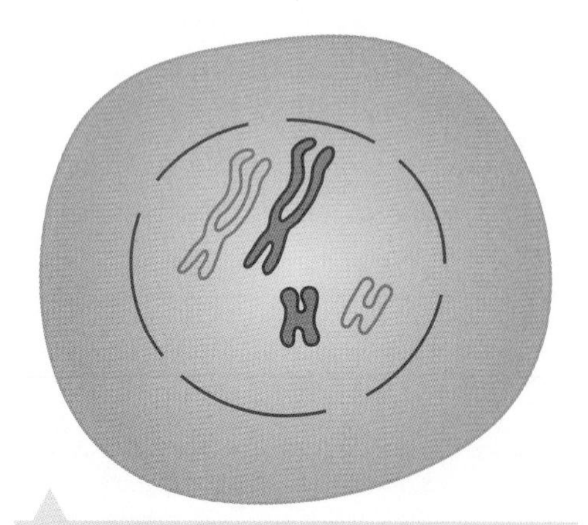

FIGURE 9.2.1 Schematic diagram of a hypothetical cell with two pairs of chromosomes; the dark chromosomes inherited from the male parent, and the light from the female parent.

1 In the space below, starting with the cell in Figure 9.2.1, draw each of the stages of meiosis, showing non-disjunction occurring with the smallest chromosome pair at metaphase I.

2 Repeat part a, but show non-disjunction occurring instead with the larger male chromosome at metaphase II.

9780170411745

Due to the huge number of genes involved, mutations caused by cell division errors generally cause dire consequences, with very few aneuploidy embryos surviving. The most common surviving aneuploids in humans are people with Down Syndrome, where three copies of chromosome number 21 are present in every body cell. This condition is also called trisomy 21.

Karyotypes are sets of chromosomes taken from cells at metaphase stage of mitosis, and arranged in order in their homologous pairs, stained and photographed.

QUESTIONS

1 In the space following, create a simplified schematic karyotype drawing of the chromosome set of a human female with Down syndrome. There is no need to try to accurately represent specific chromosome pair properties, though do attempt to illustrate their relative lengths.

2 Propose an explanation for why humans survive with trisomy 21, but not with the triploid number of chromosomes 1 or 2, for instance.

Aneuploidy can occur due to non-disjunction of sex chromosomes. Human males with Klinefelter's syndrome have two X chromosomes and one Y chromosome.

QUESTIONS

1 Create a simple karyotype drawing, in the space following, of the chromosome set of a human male with Klinefelter's syndrome.

2 Klinefelter's syndrome males can also be described as '47 XXY' individuals. Explain what the 47 refers to.

1 Mutations can occur spontaneously during:

A cell division.

B interphase between cell divisions.

C both cell divisions and/or interphase.

D the time when a cell is exposed to a mutagen.

2 Compared to mutations occurring during mitosis in an individual, the potential for effects of meiotic mutations in an individual on a whole species is:

A much less

B much greater

C the same

D unknown

3 List three differences that allow the various chromosome pairs in a cell to be distinguished from each other when examined in a karyotype.

4 Aneuploidy can result from non-disjunction in autosomal or sex chromosomes during meiosis. Klinefelter's syndrome is a sex chromosome abnormality where trisomy occurs, with affected males having 2 X chromosomes and a Y chromosome. List the different combinations of sex chromosomes in gametes that could result in XXY genotype.

5 The phenotype of an individual is determined both by its genotype and by environmental influence. This concept is often illustrated in the differences that can occur in identical or monozygous twins, that have exactly the same genotype as each other. Suggest two environmental (or lifestyle) differences that may occur between identical twins that result in phenotypic differences. Briefly describe how each difference may be brought about.

9780170411745

10 Inheritance

LEARNING

Summary

- Inheritance means acquiring genetic characteristics (genes) from parents.
- Heredity is the study of inheritance.
- Key genetics terminology:
  - homozygous genotype: two identical alleles
  - heterozygous genotype: two different alleles
  - dominant allele: only one allele necessary in genotype to produce dominant phenotype
  - recessive allele: two alleles necessary in genotype to produce recessive phenotype
  - purebreeding/homozygous parent pairs: all offspring have parental phenotype
  - parental (P) generation: original parents crossed in breeding experiment
  - first filial (F_1) generation: offspring of P generation
  - second filial (F_2) generation: offspring of F_1 generation
- A monohybrid cross is when one (mono) characteristic or trait is investigated and the P generation parents are both purebreeding but with contrasting phenotypes. F_1 generation are all heterozygous (hybrids), and F_2 generation are produced from a cross between two F_1 heterozygotes. F_2 always have a $\frac{3}{4}$ dominant : $\frac{1}{4}$ recessive phenotype ratio
- A punnett square is a simple grid tool to predict offspring genotypes and phenotypes. One parent's gamete possibilities are placed across the top of the grid, the second parent's gamete possibilities are put down the left side The four cells are then filled with the gamete combinations from above and left.
- Single gene inheritance patterns include:
  1. *autosomal dominant allele* (on non-sex chromosome). The dominant allele determines the phenotype in heterozygous genotype; two different phenotypes.
  2. *incomplete/partial dominance*. One allele does not completely dominate over the other, and heterozygous genotype has phenotype intermediate to both homozygotes; three different phenotypes.
  3. *codominance*. Both alleles are expressed fully/equally in heterozygous phenotype, which shows traits of both homozygotes; three different phenotypes.
  4. *X-linked recessive*. Phenotype is determined by a recessive allele on the X chromosome. Recessive phenotype frequencies are higher in males than females as males have only one X chromosome.
  5. *multiple alleles*. There are more than two different alleles per gene and more than three different phenotypes. For example, three different alleles result in four different human blood types: A, B, AB and O.
- Polygenic inheritance is when a trait is determined by two or more genes. There is a wide range of phenotypes with continuous variation between them. For example, human height, eye and skin colour.

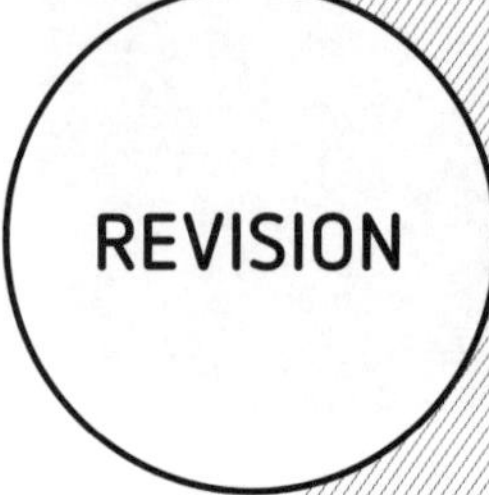

10.1 Patterns of inheritance: 'Mendel's experiments'

Briefly research Mendel's historical experiments.

1 State which type of single gene inheritance pattern we now know the seven pea plant characteristics studied by Mendel had.

____

2 Use the 'shape of the seed pod' trait to illustrate a (monohybrid) cross by:

a naming the two different seed pod shapes.

____

b stating suitable symbols to represent the dominant and recessive alleles.

____

____

c stating the genotypes of the two different purebreeding P generation plants.

Parent 1: ________

Parent 2: ________

d stating the genotype of the heterozygous F_1 generation.

____

e completing the cells of the Punnet square following with expected gamete frequencies ($\frac{1}{2}$ ________) of the two F_1 parents; expected F_2 offspring genotypes and their frequencies ($\frac{\square}{\square}$ ________); and also F_2 phenotypes (________):

F_1 CROSS (TO PREDICT F_2 GENOTYPES)		F_1 PARENT GAMETES	
		$\frac{1}{2}$ ___	$\frac{1}{2}$ ___
F_1 PARENT GAMETES	$\frac{1}{2}$ ___	$\frac{\square}{\square}$ ________ ________	$\frac{\square}{\square}$ ________ ________
	$\frac{1}{2}$ ___	$\frac{\square}{\square}$ ________ ________	$\frac{\square}{\square}$ ________ ________

3 Use part 2e to state the overall expected frequencies of the two F_2 phenotypes:

____

10.2 Patterns of inheritance: autosomal dominant alleles

In rabbits, black hair allele (B) is dominant to brown hair allele (b).

QUESTIONS

1 Draw Punnet squares to determine/illustrate expected proportions of colours in the F_1 offspring of the following parental (P) crosses.

P:	(I) BB × Bb	(II) BB × bb	(III) Bb × Bb
PUNNET SQUARES			
EXPECTED F_1 PHENOTYPIC FREQUENCIES			

2 In a certain plant, spiny seed pod allele (S) dominates over smooth pod (s). In crosses between spiny podded and smooth podded plants, numbers of offspring of each phenotype are about equal. Carry out necessary working to determine the probable genotype of each parent plant. Demonstrate that your conclusion is correct in a Punnet square showing resulting F_1 genotypes and phenotypes.

Probable parent genotypes: ________ and ________

WORKING:	PUNNET SQUARE:

10.3 Patterns of inheritance: incomplete and co-dominance

Incomplete and co-dominance inheritance patterns can be recognised by the existence of three different phenotypes for a trait, instead of the two for autosomal dominant inheritance. Conventionally, an uppercase letter representing the phenotypic characteristic is used for each allele, and a superscript, also uppercase, signifies the particular allele variation; for example: long S^LS^L; round S^RS^R; oval S^LS^R for radish shapes.

Several different genes are involved in horse coat colour determination, including the cream gene which acts to 'dilute' or lighten base coat colours expressed by other genes. This cream gene varies the base chestnut (reddish) colour to produce palomino colouring.

Alleles are partially dominant; their expression is illustrated below in Table 10.1.3.

TABLE 10.1.3 Genotypes and phenotypes for cream gene inheritance in horses

GENOTYPE	PHENOTYPE	
	NAME	DESCRIPTION
D^0D^0 (or CC)	chestnut	No dilution factor: coat has full red pigment
D^1D^1 (or CrCr)	cremello	Double dilution factor: red coat pigment diluted to pale cream; skin and eye colour also diluted; skin is pink and blue eyes common
D^0D^1 (or CCr)	palomino	Single dilution factor: red coat pigment diluted to gold, with cream to white mane and tail

9780170411745

QUESTIONS

1 List all possible matings expected to produce palomino offspring.

2 Use Punnet squares to demonstrate:

a which of the crosses from question 1 should produce the most palominos.

b the highest expected frequency of palominos from a.

PUNNET SQUARES

In a particular poultry breed, heterozygotes are described as 'erminette'. These chickens have feathers speckled with both white and black; both feather colour alleles are expressed co-dominantly. Individuals may also be homozygotes with either just black or just white feathers.

QUESTIONS

1 Choose appropriate symbols for the feather colour alleles.

Black ________

White ________

2 Write the genotypes of all three phenotypes.

Black ________

White ________

Erminette ________

3 Complete the following table to determine expected F_1 phenotypic frequencies for (A), (B) and (C):

P: PHENOTYPES	(A) BLACK × WHITE	(B) P: BLACK × ERMINETTE	(C) P: ERMINETTE × ERMINETTE
P: GENOTYPES	________ × ________	________ × ________	________ × ________
WORKING/PUNNET SQUARES:			
EXPECTED F_1 PHENOTYPE FREQUENCIES:	black: ________ white: ________ erminette: ________	black: ________ white: ________ erminette: ________	black: ________ white: ________ erminette: ________

10.4 Patterns of inheritance: sex-linked inheritance

In sex-linked inheritance, genes occur on either the X chromosome only (X-linked inheritance), or on the Y chromosome only (Y-linked inheritance). Most Y-linked genes are involved in producing male characteristics and are only passed from fathers to male offspring; always expressed. X-linked dominant inheritance is rare in humans, but X-linked recessive inheritance is more common. In males, the recessive X chromosome allele will be expressed, but there must be recessive alleles on both X chromosomes for recessive phenotype expression in females. Thus, X-linked recessive traits occur more commonly in males, including red-green colour blindness, haemophilia and a type of muscular dystrophy.

QUESTIONS

Red-green colour blindness in humans is a phenotypic condition where individuals cannot distinguish between as many colours as normal-visioned people. It is inherited via the X-linked recessive allele, X^c.

1 If the dominant allele is represented by X^C, illustrate the possible genotypes for this trait in females and in males.

____

____

2 List the genotypes that result in red-green colour blindness.

____

____

3 For each of the parental pairs below, use a Punnet square to predict frequencies of offspring phenotypes, listing males and females separately.

P:	(I) $X^C X^C$ x $X^c Y$	(II) $X^c X^c$ x $X^C Y$	(III) $X^C X^c$ x $X^c Y$	(IV) $X^C X^c$ x $X^C Y$
PUNNET SQUARES				
F_1 PHENOTYPE FREQUENCIES				

10.5 Patterns of inheritance: multiple alleles

The human A, B, O blood grouping system consists of three different alleles: I^A, I^B and i, where I^A and I^B are codominant, and i is recessive to both. Four phenotypes occur as shown in Table 10.5.1.

TABLE 10.5.1 Genotypes and phenotypes in the multiple allelic A, B, O human blood grouping system.

BLOOD GROUP PHENOTYPES	BLOOD GROUP GENOTYPES
A	I^AI^A or I^Ai
B	I^BI^B or I^Bi
AB	I^AI^B
O	ii

QUESTION

1 Parents of a baby born with O type blood suspected that they had been 'given the wrong baby', because they both had A type blood.

a What possible genotypes could the parents have?

b What possible genotypes could the O type baby have?

c Provide an illustration and an explanation to the parents, to explain why their thinking was wrong.

9780170411745

10.6 Polygenic inheritance

In polygenic inheritance, two or more (and often many) genes are involved in producing various different phenotypes across a wide range showing a smooth continuous gradation.

QUESTIONS

1 Under each of the five frequency histograms in Figure 10.6.1, place the letter that corresponds to the inheritance type illustrated, from the following list.

A Polygenic

B Multiple alleles

C X-linked recessive

D Partially or codominant

E Autosomal dominant

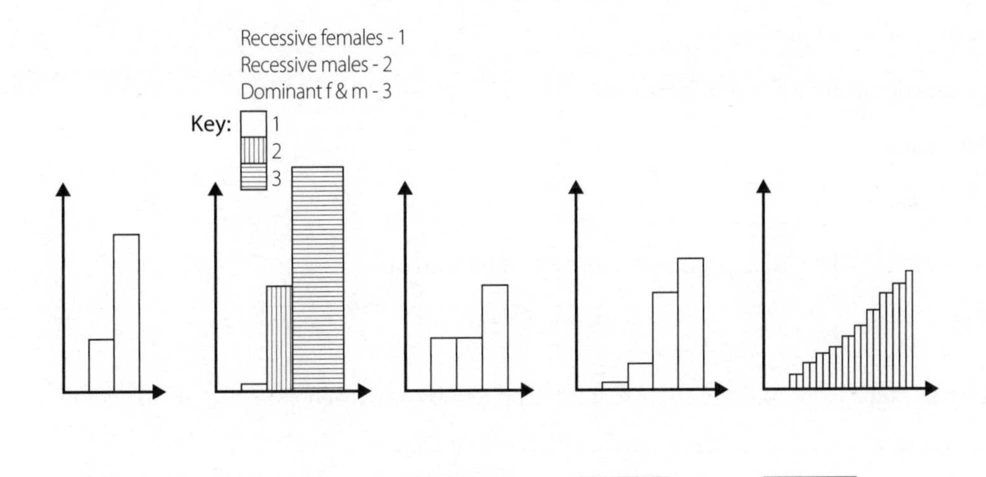

FIGURE 10.6.1 Frequency histograms: the x-axis depicts different phenotypes and the y-axis depicts approximate relative proportions of the population illustrated in each.

EVALUATION

1 Homozygous is a:

A genotype that all identical twins have.

B genotype with two alleles the same.

C genotype with two different alleles.

D phenotype that looks the same as another.

2 When three different phenotypes occur in a population, the following type of inheritance is indicated.

A Autosomal dominant alleles

B Incomplete dominance or co-dominance

C Sex-linked inheritance

D Multiple alleles

3 In rabbits, black hair allele (B) is dominant to brown hair allele (b).

P:	(i) BB × Bb	(ii) Bb × bb	(iii) Bb× Bb

a For the parental crosses indicated above, state the phenotypes of each parent in (i), (ii) and (iii):

i ______________________________

ii ______________________________

iii ______________________________

b State the parental genotypes required to ensure an all brown litter.

____

____

9780170411745

c Confirm that your answer to part b is correct using the Punnet square below.

P CROSS (TO PREDICT F_1 PHENOTYPES)		PARENT 1 GAMETES	
		$\frac{1}{2}$ ___	$\frac{1}{2}$ ___
PARENT 2 GAMETES	$\frac{1}{2}$ ___	$\frac{\square}{\square}$ ________ ________	$\frac{\square}{\square}$ ________ ________
	$\frac{1}{2}$ ___	$\frac{\square}{\square}$ ________ ________	$\frac{\square}{\square}$ ________ ________

4 Complete the following paragraph by crossing out the incorrect alternative/s in the brackets.

In X-linked inheritance of a recessive allele, there are (differences/similarities) in phenotypic frequencies between males and females. A recessive allele must be inherited from the mother for (sons/daughters/both sons and daughters) to show the recessive condition.

11 Biotechnology

LEARNING

Summary

- The term biotechnology describes the use of living things to make new products or systems. Molecular biology has revolutionised biotechnology by enabling cloning, genetic diagnosis, DNA profiling and gene sequencing.
- Restriction enzymes are enzymes isolated from bacteria that cut DNA at specific sites known as restriction sites. Sticky ends or blunt ends are formed, depending on the enzyme used.
- DNA ligase is an enzyme used to join two DNA molecules with complementary sticky ends or with blunt ends.
- Recombinant DNA technology combines DNA from different sources to generate a modified DNA sequence. Plasmids are commonly used as vectors to transport the gene of interest from an unrelated organism into bacterial cells.
- Cultures of recombinant microorganisms can generate a large number of copies of a gene of interest and pharmaceuticals like human growth hormone and insulin.
- PCR is a process through which a specific DNA sequence can be amplified for analysis. PCR is able to target a specific sequence of DNA because of its use of primers.
- Gel electrophoresis separates DNA fragments according to their size. Molecular size markers enable fragment size to be calculated.
- DNA sequencing can determine the exact nucleotide sequence of DNA fragments. Applications include forensics, metagenomics, evolutionary studies and medical diagnosis.
- DNA profiling is a technique that can be used to determine an individual's genetic profile. DNA profiling can be used to determine relatedness, identify criminals, establish or confirm pedigrees and shed light on the reproductive strategies of animals.
- The first mammal to be cloned was Dolly the sheep. The success rate of the procedure, called somatic cell nuclear transfer, was very low as it took 277 attempts and 27 pregnancies to produce one Dolly.

9780170411745

11.1 Fundamentals of biotechnology

1 Match each item in the first column with a description in the second column. Each item can only be used once.

STRUCTURE OR SUBSTANCE	DESCRIPTION OF ACTION
DNA ligase	Small circular self-replicating DNA molecule
Vector	Sorts DNA molecules based on size and charge
Primer	An enzyme that joins two segments of DNA together
Blunt ends	Specific site at which restriction enzymes cut DNA
Plasmid	An enzyme used in PCR that catalyses the synthesis of DNA
Restriction site	Vehicle to introduce DNA into a host cell
Gel electrophoresis	An enzyme that catalyses the synthesis of DNA
DNA polymerase	Result from cleavage by a restriction enzyme in the middle of the recognition sequence
Taq polymerase	Synthetic short, single-stranded DNA molecule

2 Fill in the missing words in this cloze activity.

Plasmids are extracted from ______________ by rupturing the cell membranes and cell walls. Similarly, the ______________ of interest is isolated from the ______________ organism. The same ______________ enzyme is used to cut the ______________ DNA and the DNA of the ______________ to be inserted, to ensure they have ______________ sticky ends. The plasmid ______________ and the gene of interest are ______________ together and their ______________ ends pair. DNA ______________ is used to join the two segments to form ______________ plasmids. These plasmids are added to a ______________ culture, where they are taken up by some ______________ in a process called ______________. When the bacteria reproduce by dividing, the ______________ is also ______________. This generates numerous ______________ of the recombinant DNA. A process called ______________ selection can be used to identify ______________ bacteria.

3 Complete the table below, which summarises PCR, by filling in the blank spaces. Use numbers 1-3 in the appropriate column to show the order in which these steps occur.

NAME OF PROCESS	CONDITIONS FOR PROCESS	ORDER	PURPOSE
			New DNA strands are synthesised starting from primers
	Temperature raised to 95°C		
Annealing			

11.2 Using gel electrophoresis for genetic testing

Genetic testing is used to determine whether a person possesses an allele for a gene that is associated with a specific disease. If the gene mutation changes a restriction site, it can be detected by gel electrophoresis, using a technique known as restriction fragment length polymorphism (RFLP) analysis. Sickle cell anaemia can be investigated in this way.

The allele for sickle cell anaemia is due to a point mutation, which causes a change in a single amino acid in the beta-globin molecule of haemoglobin. A single base-pair change (from CCTGAGG to CCTGTGG) causes sickle cell anaemia. This single base change results in the loss of one of the restriction sites of the restriction enzyme MstII. As a result, MstII digestion of DNA in the region of the mutation yields two fragments in a normal patient (1150 bp and 200 bp), but only one in an affected patient (1350 bp). These different sized fragments are detected by gel electrophoresis (Figure 11.2.1).

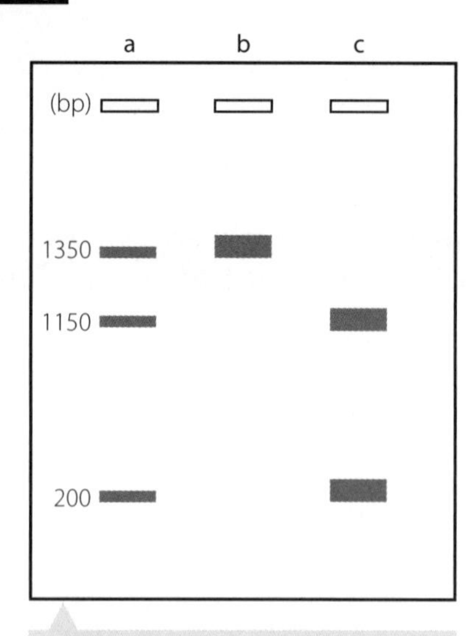

FIGURE 11.2.1 The electrophoresis gel shows the patterns of individuals: with sickle cell anaemia, without sickle cell anaemia and a carrier.

QUESTIONS

1 Describe the function of a restriction enzyme.

____

2 Describe sticky ends in relation to a restriction enzyme.

____

____

3 Describe the role of gel electrophoresis in biotechnology and discuss the role of the agarose gel in this process.

____

____

____

____

____

4 Describe one way in which DNA is made visible in gel electrophoresis.

____

____

____

5 Identify the lanes in Figure 11.2.1 (a, b or c) that show the results for:

a an individual with sickle cell anaemia. ______

b an individual without sickle cell anaemia. ______

c a carrier of sickle cell anaemia. ______

11.3 DNA profiling in the study of reproductive strategies

Family size is highly variable in Black Swans and varies between one and seven. Family size distribution seems to be bimodal with most families containing 1–3 cygnets or 5–7 cygnets. This has led to speculation that larger families are the result of brood parasitism, when a female lays her eggs in the nest of a second female and leaves this second female to raise her young. This process is quite common in ducks, but has not been investigated in Black Swans.

A DNA profile for both the mother and her cygnet will be used to determine if the female is the biological mother. Your aim is to determine, using DNA profiling, whether brood parasitism occurs in Black Swans and whether this explains the larger number of cygnets in some families.

DNA was obtained by capturing swans, collecting a small blood sample from each, and extracting the DNA. Five STRs were analysed (Cam 1, Cam 2, Cam 3, Cam 4 and Cam 5). Using PCR, these five regions were amplified in all adults and cygnets of eight families of swans then separated using agarose gel electrophoresis. Figure 11.3.1 shows the resulting gel. Each individual has two alleles for each STR, but sometimes only one band is observed as the individual has two identical alleles.

QUESTIONS

1. Compare the profile of the mother of each family, and the profile of each cygnet, and determine if the female could have been the biological mother of the cygnet.
2. Record your results in the second column of Table 11.3.1.
3. Calculate the proportion of parasitic cygnets in each family and include these in the third column.
4. Determine if there is any difference in the proportion of parasitic cygnets between small and large families.

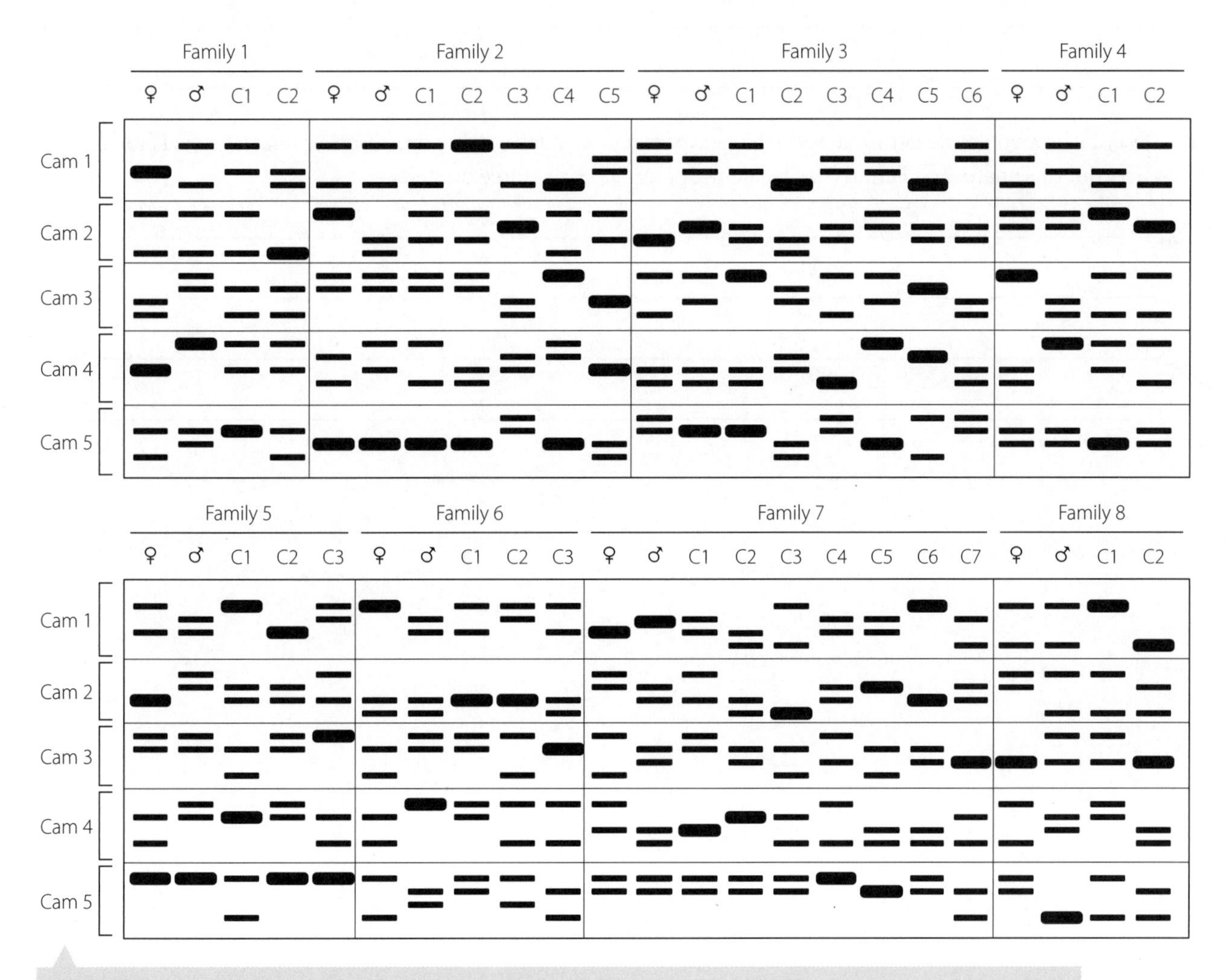

FIGURE 11.3.1 DNA profiling for the eight black swan families that include a social mother (♀), social father (♂) and cygnets (C)

TABLE 11.3.1

FAMILY	BIOLOGICAL CYGNETS	PARASITIC CYGNETS
1		
2		
3		
4		
5		
6		
7		
8		

5 Using your results identify any evidence of brood parasitism in Black Swans.

6 Calculate the maximum proportion of parasitic cygnets in this sample.

7 Explain whether the results confirm the belief that large Black Swan families are due to brood parasitism.

8 Describe how you could determine whether a cygnet has been fathered by a male other than its social father. Carry out this analysis on Family 5 to decide the paternity of the three cygnets.

1 Which one of the following combinations correctly matches a substance or process, its mode of action and its purpose?

SUBSTANCE OR PROCESS	MODE OF ACTION	PURPOSE
A DNA ligase	Joining of two fragments of DNA	Identify individuals with a genetic disease
B Restriction endonuclease	Cutting of DNA into similar lengths	Generate matching sticky ends when making recombinant DNA
C Gel electrophoresis	Separating fragments of DNA	Identify individuals with a genetic disease
D DNA profiling	Aligning fragments of DNA	Identify individuals in natural disasters and plane crashes

2 Which of the following steps is not part of making recombinant DNA?

A DNA ligase is used to join the two segments of DNA to form recombinant plasmids.

B Plasmids are extracted from bacteria by rupturing the cell membranes and cell walls.

C DNA polymerase is used to amplify the foreign gene before it is inserted into the plasmid.

D The plasmid vectors and the foreign gene are mixed together and their sticky ends pair.

3 In the process of DNA profiling, which of the following statements is correct?

A The agarose gel separates segments of DNA according to size, with smaller fragments at the top of the gel, near the sample wells, and larger fragments at the bottom.

B The polymerase chain reaction is used to amplify the DNA of up to 20 different DNA regions in the sample.

C In the gel electrophoresis stage, overheating of the gel can be a problem because the restriction enzymes may denature.

D An advantage of the polymerase chain reaction is that important regions of the DNA that code for differences like eye colour can be amplified.

4 In the polymerase chain reaction, a solution containing DNA and an enzyme is heated and cooled. The solution is cooled to:

A allow the enzyme to function at its optimum temperature.

B break the weak bonds between the bases in the DNA.

C allow complementary primers to copy the DNA.

D ensure the DNA is not denatured.

5 Predict the minimum band sharing percentages in the DNA profiles of a mother and her baby. Explain your answer.

6 When conducting PCR, some unwanted DNA molecules are sometimes present.

a Identify the possible consequences of having an unwanted DNA molecule in the PCR.

b Identify two possible sources of this contamination.

c Suggest what could be done to prevent this contamination from occurring.

9780170411745

12 Evolution

LEARNING

Summary

- Darwin's theory of evolution by natural selection proposed that the vast diversity of organisms living today descended from a common ancestor. He showed that changes within species leads eventually to the development of new species.
- Microevolution results in a change in the frequency of various alleles within a population, in a process called natural selection.
- Macroevolution describes major evolutionary changes above the level of species and results in the divergence of taxonomic groups, in which the descendant is in a different taxonomic group to the ancestor.
- Earth and its inhabitants have changed enormously over 3.5 billion years.
- The positions of landmasses are in constant change. Geological and fossil evidence tells us that 200 mya a single supercontinent – Pangaea – existed, which would later separate into smaller landmasses.
- Earth's climate has oscillated between hot, humid periods and cold, dry periods. Some of these changes were rapid and dramatic, causing major changes to sea levels and temperatures; others occurred more slowly over time.
- The fossil record shows that periods of evolutionary radiation, where many new species have evolved from a single ancestral form, have always followed mass extinctions. There have been five major extinction events in the history of life on Earth.
- Different organisms share both structural and molecular similarities. Examination of the genes of different organisms indicates that all modern life descended from a single population of organisms.
- Comparative genomics provides evidence for the theory of evolution and helps us map the degree of species relatedness. The closer the relationship between two organisms, the greater the similarities between their DNA.

12.1 Describing evolution

Communication is vital when discussing and explaining the evolution of life on Earth.

Show your understanding by writing the correct concept from the list below into Table 12.1.1, to match its description in the right column.

Comparative genomics	Phylogenetic relationships	Molecular homology
Molecular phylogeny	Macroevolution	Evolutionary radiation
Microevolution		

TABLE 12.1.1

	CONCEPT	DEFINITION
1		evolutionary relationships that exist between individuals or groups of organisms
2		the process of contrasting DNA sequences in different organisms
3		when the common ancestry of organisms is identified from shared biomolecular elements used to test their relationships
4		small scale variation of allele frequencies within a species or population, in which the descendant is of the same taxonomic group as the ancestor
5		the study the evolutionary relationships between organisms by using DNA data
6		an increase in taxonomic diversity or morphological disparity over time
7		variation of allele frequencies at or above the level of species resulting in the descendant being in a different taxonomic group to the ancestor

12.2 Describing evolution

Both mitochondrial DNA and chloroplast DNA have been useful in the determination of relationships between related organisms. The data in Table 12.2.1 shows the differences in the mitochondrial DNA nucleotide sequences between pairs of vertebrates.

TABLE 12.2.1 Comparison of percentage differences in DNA nucleotide sequences between pairs of vertebrates

GROUPS COMPARED	DIFFERENCE IN DNA SEQUENCES
Human – chimpanzee	1.6%
Human – gibbon	3.5%
Human – rhesus monkey	5.5%
Human – African galago	28.0%
House mouse – Norwegian rat	20.0%
Cow – sheep	7.5%
Cow – pig	20.0%

9780170411745

QUESTIONS

1 Recall the form and function of DNA in mitochondria and chloroplasts.

2 State two ways in which next generation DNA sequencing methods differ from those used in the Human Genome Project over 20 years ago.

3 What does the degree of similarity of the nucleotide sequences of living things tell us about their relationships?

4 Compare the similarity in the genomes of the cow and sheep with that of the cow and the pig. What can you conclude from this?

5 Consider the degree of similarity between the DNA of humans and the four other primates – chimpanzee, gibbon, rhesus monkey and the African galago (bushbaby). Discuss the evolutionary relationships in these primates.

6 Relate the results in Table 12.2.1 to the concept of macroevolution.

12.3 Case study: what did terror birds eat?

Two-metre-tall prehistoric 'terror birds' belonging to the genus *Gastornis*, looked so fierce that palaeontologists assumed they were terrifying predators. New research finds that the would-be carnivores were probably herbivores.

The flightless terror bird lived in what is now Europe between 55 and 40 mya. This was after the extinction of dinosaurs in the Cretaceous Period and at a time when mammals were at an early stage of evolution and relatively small. The terror bird was thought to have been a top predator at that time, using its huge, sharp beak to grab and break the neck of its prey.

Although biomechanical modelling of its bite force supported this theory, footprints left behind by an American cousin of *Gastornis* do not show imprints of sharp claws, a feature needed to grapple prey. Today's raptors, for example, sport such sharp claws. Another clue is more obvious — the bird's hefty size and build. That bulk would not make for a very swift hunter.

To explore its diet, researchers took a geochemical approach. They analysed the fossilised bones of the birds, focusing on calcium isotope composition. Isotopes are atoms of the same element with different numbers of neutrons. The team showed that lighter isotopes of calcium become progressively enriched along a food chain. This means that the proportion of lighter to heavy isotopes found in bones of carnivores is greater than those of herbivores.

The scientists tested the method with herbivorous and carnivorous dinosaurs — including top predator T. rex — as well as mammals living today. They discovered that the calcium isotope compositions of *Gastornis* bones are similar to those of herbivorous mammals and dinosaurs, and not to carnivorous ones.

Even if the food was just plant based, it had to have been large and tough, given the impressive beaks of these amazing terror birds.

Source: Viegas, J. (2013) '"Terror bird" was scary-looking vegetarian', Discovery News online, 29 August.

QUESTIONS

1 Suggest why it had been assumed that *Gastornis* was a predator, and state two pieces of evidence that pointed away from this before the so-called 'geochemical approach'.

2 Explain why mammals, birds and flowering plants underwent evolutionary radiation after the extinction of the dinosaurs.

3 Suggest an alternative reason for the apparent absence of raptor-like toe claws on *Gastornis*.

4 Make a prediction, giving reasons, of the approximate calcium isotope composition of omnivores, that consume both plant and animal material.

5 If *Gastornis* was herbivorous, suggest the types of plants it might have eaten.

9780170411745

12.4 Case study: the evolution of the marsupials

Marsupials are a type of mammal that carry their young in a pouch. They are found only in Australasia and South America and include kangaroos, koalas, wombats, possums and opossums. DNA evidence supports a South American origin for marsupials, with Australian marsupials arising from a single Gondwanan migration of marsupials from South America to Australia, via Antarctica.

The evolutionary relationships among the seven marsupial orders – four Australasian and three South American – have not yet been resolved. The two recently sequenced marsupial genomes of the South American opossum and the Australian tammar wallaby, provided scientists with an opportunity to apply a completely new approach to resolve marsupial relationships.

The exhaustive computational and experimental evidence provided important insight into their evolution and resolved most branches of the marsupial evolutionary tree. The results of this study are shown in Figure 12.3.1.

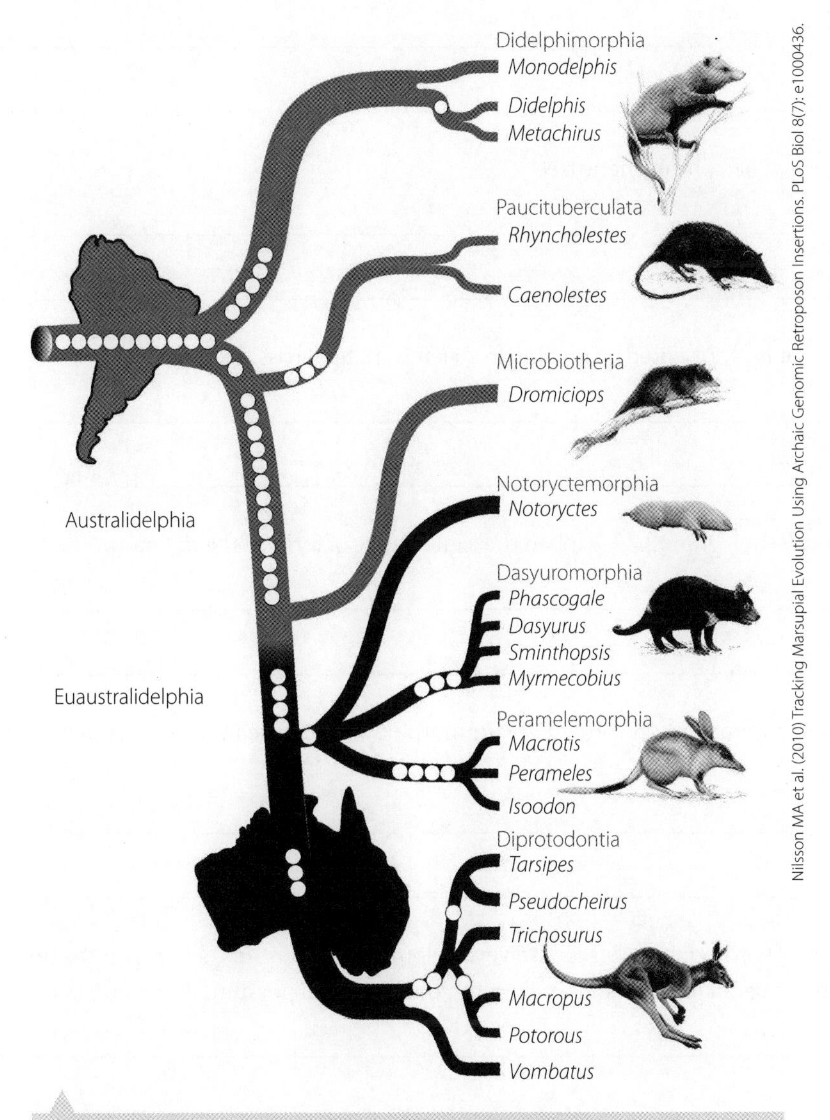

FIGURE 12.3.1 Diagram showing evolutionary relationships between the marsupials. The first marsupials arose in South America, then migrated to Australia, via Antarctica.

QUESTIONS

1 Explain how the marsupial ancestor was able to move from South America to Australia via Antarctica, around 60 mya.

2 Would the type of evolution shown in Figure 12.3.1 be an example of microevolution or macroevolution? Give reasons for your answer.

3 Describe the purpose of a phylogenetic tree.

4 State the factor that is represented by moving from left to right across the diagram.

5 Using an example from Figure 12.3.1, explain the significance of forks in the diagram.

6 Predict which other member of the order Dasyuromorphia would have DNA most similar to the genus *Dasyurus*, the quoll.

7 Consider the order Diprotodontia. Discuss the evolutionary relationships of *Tarsipes*, the honey possum, *Pseudocheirus*, the ring tail possum and *Trichosurus*, the brush tail possum.

8 Explain why this study could be called comparative genomics.

9 Describe the importance of large scale and exhaustive computations in a study of this kind.

About 25 mya, a combination of ice ages and a drying out of inland northern Australia brought about the destruction of the existing tropical habitat. Lush rainforests were replaced by arid grasslands. Leaf-eating diprotodonts and the huge range of possums could not survive with the limited food available.

The impact of the shifts in habitat that the Australian continent has experienced is summarised in the evolutionary tree of Australian mammals shown in Figure 12.5.1

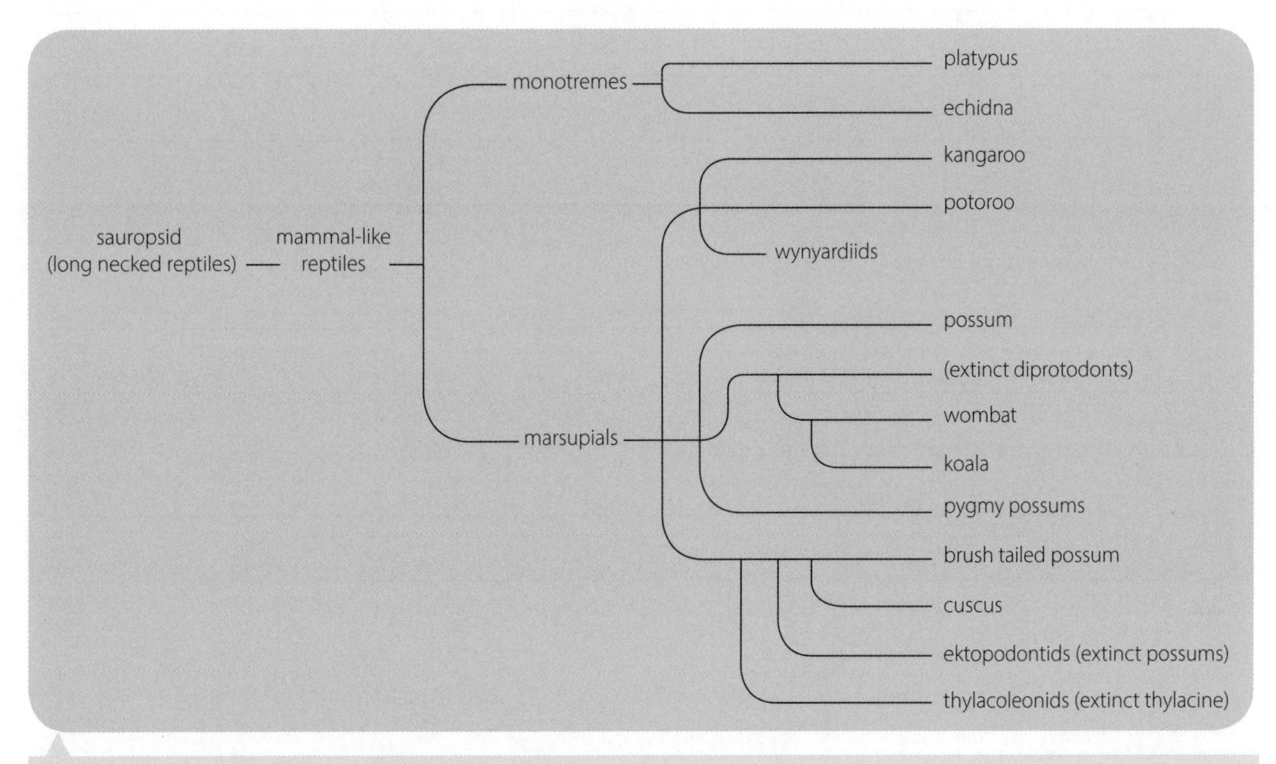

FIGURE 12.5.1 The evolutionary relationships of Australian marsupials and monotremes.

1 Use Figure 12.5.1 to answer the following questions.

a State two ways in which scientists could have collected the evidence used to determine the evolutionary relationships shown.

____

____

____

b The diagram shows several groups became extinct when the forests of northern Australia were replaced by grasslands. Using this example, explain why major extinction events are followed by evolutionary radiation.

____

____

____

c Scientists assert that the evolutionary relationships of Australian marsupials and monotremes, as shown in Figure 12.5.1, provide strong support for Darwin's concept of descent with modification. Discuss this assertion.

d Use the diagram to name the closest relatives of the extinct diprotodonts.

2 Compare and contrast Lamarck's and Darwin's theories of evolution.

3 Distinguish between microevolution and macroevolution.

4 Explain why comparative genomics provides evidence for the theory of evolution.

9780170411745

13 Natural selection and microevolution

LEARNING

Summary

- A biological population is made up of a group of individuals of the same species that live in the same geographic area and readily interbreed to produce fertile offspring.
- Variations exist in populations. The total collection of gene variations – alleles – within a population make up the gene pool.
- The allele frequency is the frequency of occurrence or proportions of different alleles of a particular gene in the total collection of alleles within a population.
- The frequency of alleles in a population can be affected by mutation, immigration, emigration and reproduction rate. Mutation is the only source of new alleles.
- Gene flow occurs when individuals migrate into or out of a population.
- Genetic drift applies to random changes in small populations. There is a chance that some alleles present in a parental group will not be passed on to offspring. These alleles may be permanently lost from the gene pool.
- A bottleneck effect occurs when a large population is suddenly reduced. Alleles present in the smaller population may not be representative of the larger population.
- The founder effect occurs when a small group of individuals migrates and establishes a population in a new location. The isolated population has less genetic diversity than the original population.
- Natural selection acts on variations in the phenotypes of individuals so that some survive and reproduce while others do not. The population evolves over time.
- Selection pressures are factors that influence the survival and ability of an individual to reproduce within a population.
- When the environment changes very little, natural selection is said to be stabilising. Changes in the environment lead to selective pressures causing directional or disruptive selection.

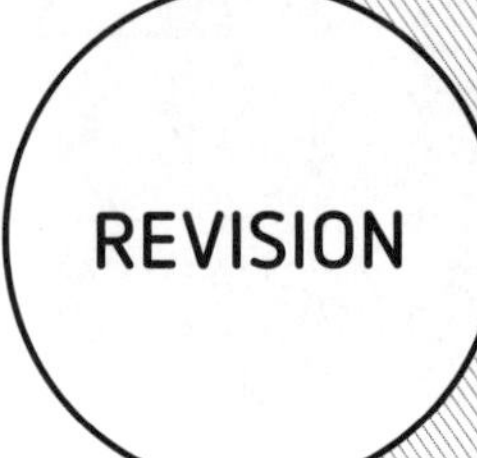

13.1 Gene pools

The coat colour of a population of mice in the hay shed of a farmer is determined by a gene that has two possible alleles – brown and white. Brown colour and white colour are co-dominant.

Shutterstock Premier/PETERVONWIEN/Shutterstock.com

FIGURE 13.1.1 White and brown coloured mice

The total number of mice in the hay shed is 160. Of these, 90 are light brown, 55 are dark brown and 15 are white.

QUESTIONS

1 Define the biological term 'population'.

____

____

2 Define 'gene pool'.

____

3 Assign allele symbols for the mouse coat colour alleles.

____

4 Assign genotypes to the phenotypes of the mice.

a Dark brown mouse ______________________________

b Light brown mouse ______________________________

c White mouse ______________________________

5 Calculate the total number of alleles for coat colour in the gene pool.

____

6 Calculate the frequency of:

a brown alleles

____

b white alleles

____

7 Use figure 13.1.1 to explain the different allele frequencies.

____

____

8 Describe where new alleles come from.

____

13.2 Changes to allele frequencies

The allele frequency in a population can be affected by mutation, gene flow, reproductive rates and genetic drift.

QUESTIONS

1 Calculate the frequencies of the B and b alleles in the population sample shown in figure 13.2.1

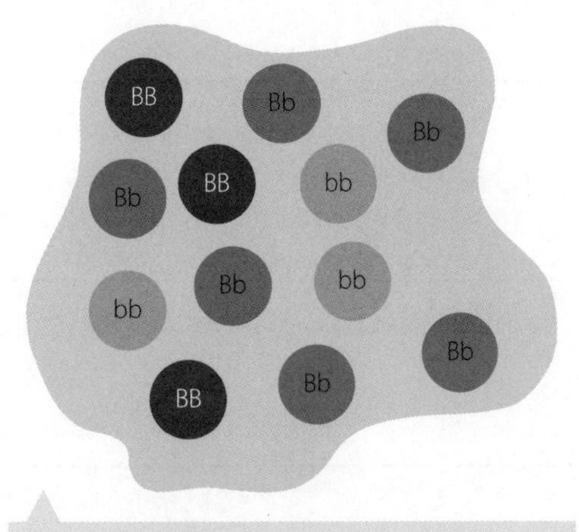

FIGURE 13.2.1 Allele population sample

B alleles: ____

b alleles: ____

2 Calculate the allele frequencies if:

a three BB individuals migrated into the population.

B alleles: ____

b alleles: ____

b three bb individuals left the population.

B alleles: ____

b alleles: ____

3 Figure 13.2.1 illustrates a small population. Compare the magnitude of allele frequency change that would occur in this small population if three white individuals left, with the change in allele frequencies if three dark mice migrated into the population of hay shed mice (from Activity 13.1). Make a statement about the impact of migration on allele frequencies in different sized populations.

____

____

____

4 Allele frequencies can change by chance, particularly in small gene pools. Genetic drift, including the bottleneck effect and the founder effect are all factors that alter allele frequencies. Describe and create a graphic representation for each of the factors.

a Genetic drift

Description:

____

____

Graphic representation:

b Bottleneck effect

Description:

____

____

Graphic representation:

c Founder effect

Description:

Graphic representation:

13.3 Natural selection

Natural selection acts on individuals to produce changes in whole populations over time. Natural selection acts on the phenotypes of individuals, so that some survive and reproduce while others do not.

1 Explain why genetic variation is necessary for natural selection to act on.

2 Figure 13.3.1 is a diagrammatic representation of natural selection. In this example, the darker alleles confer an advantage over lighter ones. Explain what may be happening at each step (level) of the diagram and explain your reasoning. Respond in the space at the right of the diagram.

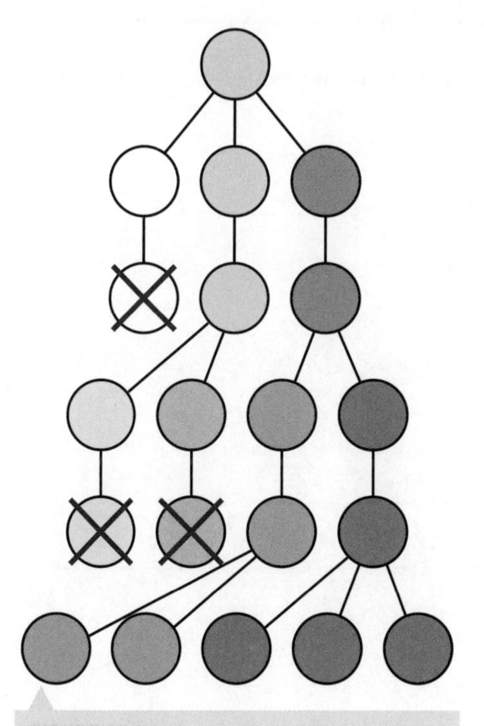

FIGURE 13.3.1 Diagrammatic representation of natural selection

3 Illustrate each of the steps set out in Figure 13.3.1 by using the example of the peppered moth, *Biston betularia* at the time of the Industrial Revolution in Britain.

4 Natural selection is most obvious when it is leading to changes in the gene pool of a population, causing some observable change in phenotype. Explain the following forms of phenotypic selection.

a Stabilising selection

b Directional selection

c Disruptive selection

5 Which type of phenotypic selection is being illustrated in each of the graphs below? Explain how you came to your conclusion.

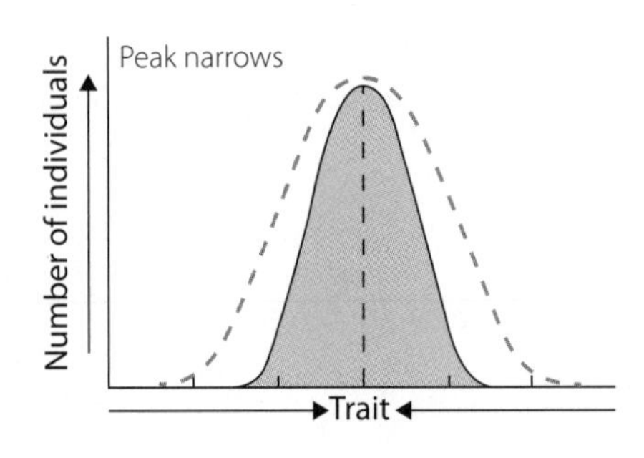

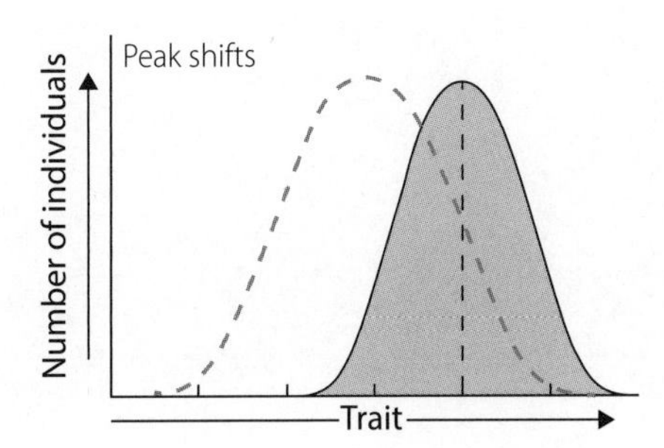

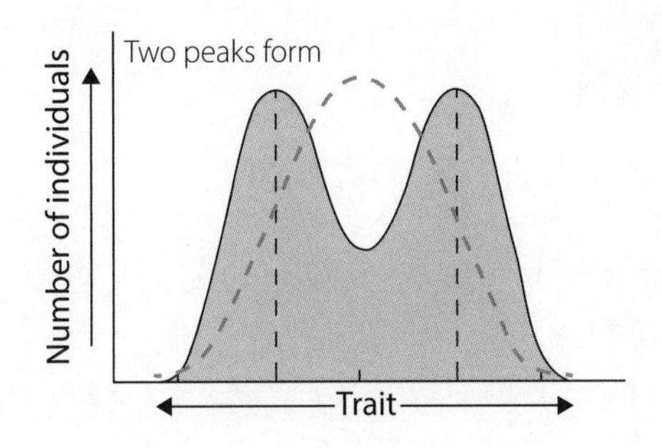

6 Suppose that the hay shed in Activity 13.1 was almost empty. The walls and floors of the shed are white in colour.

a Predict what would happen to allele frequencies as a result of natural selection. Explain your reasoning.

b Describe the type of phenotypic selection in this example.

1 A scientist collected the genotypes of a small population of fish over several generations. The presence of stripes is determined by the S allele, while unstriped fish can only have the ss genotype. Results are shown in the table below.

	GENOTYPE NUMBERS IN GENE POOL		
	GENERATION 1	GENERATION 2	GENERATION 3
SS	10	7	2
Ss	30	20	10
ss	10	23	38
Total fish	50	50	50

a Explain what is meant by 'allele frequency'.

____

b Calculate the total number of alleles in the gene pool.

____

c Calculate the allele frequency for the s allele in generations 1, 2 and 3. Show your working out.

____

____

____

d Name and describe the likely process that caused the change in allele frequency.

____

____

e Name the term applied to the type of phenotypic selection in this example. Explain what is meant by that term.

____

____

2 Retinitis pigmentosa is an inherited disease causing progressive blindness in humans. The frequency of the allele causing retinitis pigmentosa is four times greater in the population of an island in the Atlantic Ocean compared to the original British population that colonised it.

a Name and describe the process that most likely caused this change in allele frequency.

____

____

 9780170411745

b There was a natural disaster on this island after colonisation. Discuss how this could influence your answer to part a.

---

3 Myxoma virus was introduced in Australia in the early 1950s as a biological agent to kill rabbits. After first being introduced, rabbit numbers fell by 80% in some areas. In the years that followed, rabbit numbers began to increase. The number of rabbits resistant to the myxoma virus increased from around 5% in the early 1950s to close to 90% by the late 1950s. Explain how natural selection would bring about the increase in myxoma resistant rabbits.

---

4 'Natural selection acts on some individuals developing mutations that suit them better to their environment compared to other members of the population.'

Identify one part of the statement that is correct and identify another part of the statement that is incorrect. Rewrite a correct statement.

Correct part of statement:

Incorrect part of statement:

Correct statement:

---

14 Speciation and macroevolution

Summary

- Groups of actual, or potentially, interbreeding natural populations that are reproductively isolated from others are defined as a biological species.
- The morphological species concept is used to identify different species that can only be observed in the fossil record based on their physical and physiological characteristics.
- New species form when a single population becomes separate populations that are unable to interbreed due to changes that produce physical, biological or behavioural barriers. Different selection pressures act on the separated populations. Lack of gene flow leads to reproductive isolation.
- Pre-reproductive isolating mechanisms prevent organisms from being able to interact to reproduce. They include geographical, temporal, behavioural and morphological mechanisms.
- Post-reproductive isolating mechanisms do not prevent mating from occurring but they do prevent young from being produced. They include gamete mortality, zygote mortality and hybrid sterility.
- In allopatric speciation, gene flow between populations is disrupted as they become physically separated through geographic isolation. The populations diverge because of different selection pressures. Habitat fragmentation is an example of allopatric speciation.
- Species sometimes diverge without physical isolation. This is termed sympatric speciation.
- Parapatric speciation occurs when populations are separated by an extreme change in habitat. Populations may interbreed in bordering areas.
- Divergent evolution is a pattern of evolution where differences between groups of organisms accumulate to a critical point that leads to speciation.
- Convergent evolution is a pattern that occurs when unrelated organisms evolve similar adaptations in response to their environment.
- Parallel evolution describes the pattern where unrelated species in the same environment evolve independently of each other and develop similar characteristics.
- Coevolution is a process whereby an evolutionary change in one species has affected an evolutionary change in another species.
- Changes in the environment of an organism may make the habitat so unsuitable that all members of the species die and the species becomes extinct.

 9780170411745

14.1 Mechanisms of isolation in speciation

A 95 million year old dinosaur bone was first found near Winton, 177 km north-west of Longreach in Queensland in 2005. The announcement of the discovery of a new species of dinosaur named *Savannasaurus elliottorum* in 2016 followed years of work to release the rest of the bones from the surrounding soil. The dinosaur's ancestors probably came from South America.

QUESTIONS

1 Define the meaning of biological species.

____

2 The identification of *Savannasaurus elliottorum* as a new species would have been based on the morphological species concept. Explain why this was used rather than the concept of biological species.

____

____

3 Outline what is likely to have happened for this new species to evolve. Include the following terms in your answer – gene flow, gene pool, selection pressure and reproductive isolation.

____

____

____

4 Predict how a region with constant environmental conditions would impact on evolution.

____

5 The key to the formation of new species involves reproductive isolation combined with selection pressures, leading to a disruption of the flow of genes. Isolating mechanisms separate two groups and prevent them from producing fertile, viable offspring. Distinguish between pre-reproductive isolating mechanisms and post-reproductive isolating mechanisms.

____

____

6 Describe the following pre-reproductive mechanisms of isolation.

a Geographic

____

b Temporal

____

c Behavioural

______

d Morphological

______

e Spatial

______

7 Describe the following post-reproductive mechanisms of isolation.

a Gamete mortality

______

b Zygote mortality

______

c Hybrid sterility

______

8 Complete Table 14.1.1 by matching up the type of mechanism of isolation listed in questions 6 and 7 to the example.

TABLE 14.1.1

EXAMPLE	TYPE OF ISOLATING MECHANISM
A horse and donkey produce an infertile mule	
Variation in an animal species size prevents the largest in the species mating with the smallest	
A massive earth quake creates a mountain range separating populations from each other	
A population of snakes emerges from hibernation at a different time to the main population	
Coral forms in a cooler area of a reef	
Pollination and fertilisation of flowers is successful but no seeds develop	
Plants flower at different times of the year	
The mating call pattern of a frog changes	
A new river flow forms	
Reproductive structure shape of snails differs	
A species of rice pollinates another species of rice that produces seeds that grow into plants that are sterile	
Birds differ in their mating rituals	
Gametes no longer fuse together	

9780170411745

14.2 Modes of speciation

Gene flow can be interrupted between populations of existing species. The populations may be geographically isolated by a physical barrier or sometimes species diverge without any physical or geographic isolation.

QUESTIONS

1 Describe the following terms and provide one example of each.

a Allopatric speciation

Description: ____________________

____

Example: ____________________

____

b Sympatric speciation

Description: ____________________

____

Example: ____________________

____

c Parapatric speciation

Description: ____________________

____

Example: ____________________

____

2 The loss of habitat caused by agriculture as well as other human activities such as urban development, mining and pollution has caused habitat fragmentation in many areas.

a Explain what is meant by habitat fragmentation.

____

b Outline the steps in speciation after populations are cut off from each other due to habitat fragmentation.

____

____

c Describe one way to overcome habitat fragmentation.

____

____

3 The modes of speciation can be shown in a graphical representation. Place the following graphics in the correct spaces in Table 14.2.1. A graphic can be used in more than one space.

TABLE 14.2.1

	ALLOPATRIC SPECIATION	SYMPATRIC SPECIATION	PARAPATRIC SPECIATION
Original population			
First step of speciation			
Reproductive isolation			
New species			

14.3 Patterns of evolution

Diversification of species over time can follow several different patterns. Selection pressures can have different effects on the ways in which species exposed to them evolve. Evolution gives rise to groups of organisms that become very different from each other. Evolution can also give rise to groups of organisms that are similar yet not related. Four patterns of evolution are identified: divergent, convergent, parallel and coevolution.

QUESTIONS

1 Write a brief description of each term.

a Divergent evolution: ______________________________

____

b Convergent evolution: ______________________________

____

c Parallel evolution: ______________________________

____

d Coevolution: ______________________________

____

2 Draw a diagram that represents:

a divergent evolution **b** convergent evolution **c** parallel evolution

3 Explain why Australian marsupial dunnarts and mice have a similar body plan even though they do not share common ancestry.

____

____

4 Tasmanian devils (ground-dwelling carnivores) and marsupial moles (dune-burrowing insectivores) are related because they have a common marsupial ancestor. However, they show quite different feeding structures that adapt them to different diets. Identify and explain the pattern of evolution that caused this diversification.

____

____

5 Sea snails are the natural prey of crabs. Through natural selection, crabs develop more powerful claws to pierce the snail's shell. In response snails with thicker shells are favoured by natural selection.

a Identify this pattern of evolution.

____

b Provide an example of this pattern of evolution that occurs in beneficial relationships.

____

6 Interpret the drawings in Figure 14.3.1. Identify the pattern of evolution and explain why you chose that pattern in the lines below.

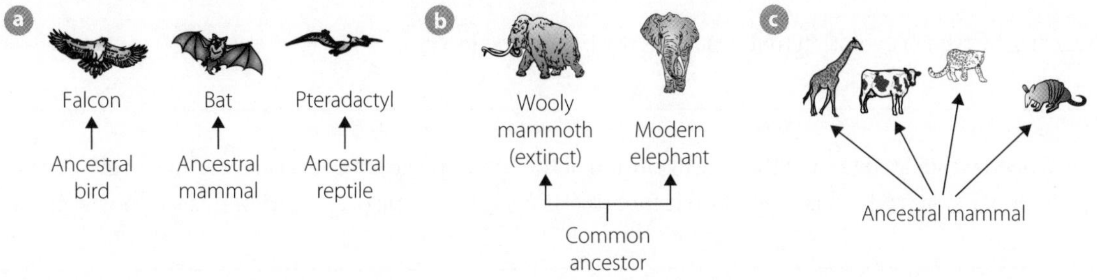

FIGURE 14.3.1 Various patterns of evolution

7 Pterodactyls and woolly mammoths, shown in figure 14.3.1, became extinct. Explain how a population bottleneck often leads to extinction of a species.

____

____

____

1 More than fifty new species of spiders were identified by a team of scientists, rangers and traditional owners on the Cape York Peninsula in 2017. Describe what evidence the group needed to substantiate claims that the spiders belong to new species.

2 Three different species of bats are found in an area of Queensland tropical rainforest. They have evolved from a common ancestor. They are similar in appearance but emit different frequencies of echolocation signals. One of the bat species feeds on insects close to the forest edge, another feeds on insects above the trees and the third bat species feeds on nectar within the forest.

a Suggest the type of evolution that describes the bat diversification.

b It has been wrongly suggested that the evolution of the three species of bats is an example of allopatric speciation. Explain why this suggestion is incorrect. Suggest the type of speciation this is an example of.

3 Explain why the survival of a population is threatened when there is low genetic diversity.

4 Name one pre-reproductive isolating mechanism involved in speciation.

5 Explain how isolation may result in speciation.

9780170411745

6 Name and describe what type of evolution these diagrams represent.

a

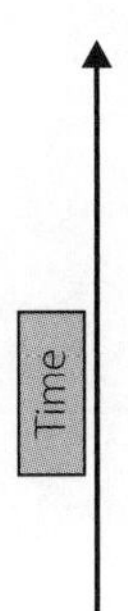

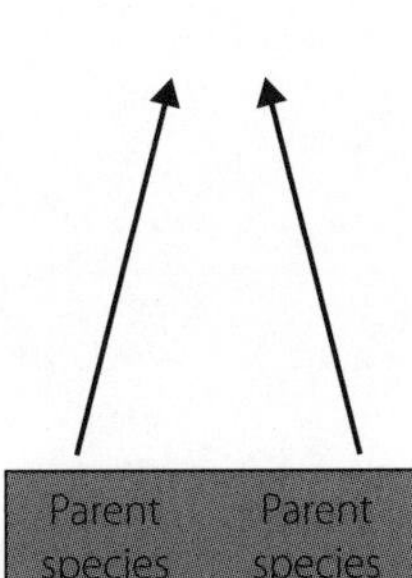

b

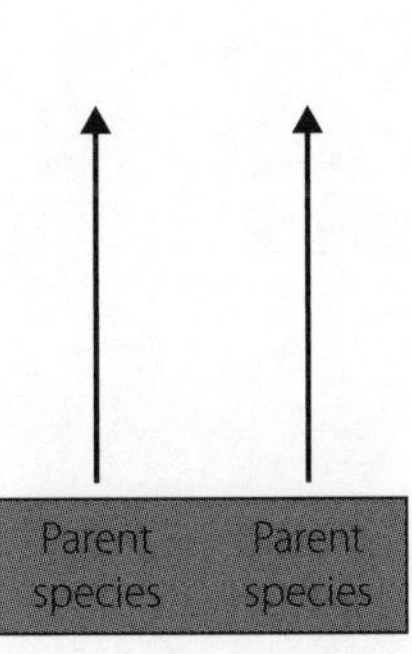

7 Common garter snakes are able to eat toxin-producing newts. Over time, the snakes have become resistant to the newt toxins. Those newts producing more toxins are able to survive. This evolutionary change in the newts will again trigger the garter snakes to evolve resistance to higher levels of toxins.

a Explain why scientists regard this as an example of predator-prey coevolution.

b Give an example of mutualistic (both benefit) coevolution.

8 Describe two key differences between allopatric and sympatric speciation.

BIOLOGY UNITS 3 & 4

MULTIPLE-CHOICE QUESTIONS

1 Predation, competition and climate

A are examples of limiting factors.

B can all be used to classify members of a species.

C are constant within and between ecosystems.

D have no effect on biodiversity.

2 The following diagram shows the evolutionary relationships between a number of species.

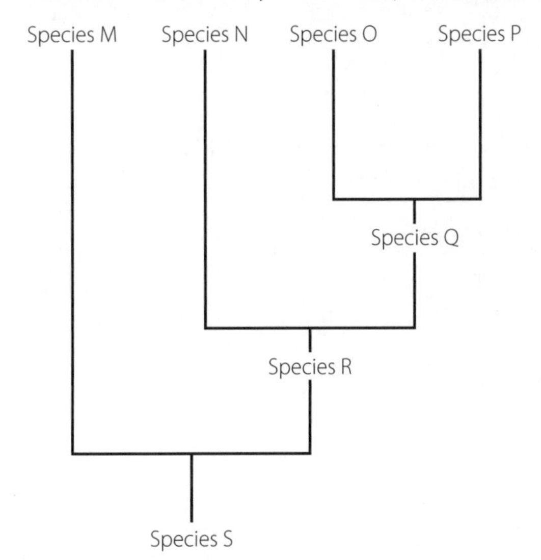

The diagram indicates that

A species M and N are the most closely related species shown.

B species Q is the most recent common ancestor of species O and P.

C species R is a common ancestor to species M, N and O.

D species S is the most recent common ancestor of species O and P.

3 Six sailors were shipwrecked on a desert island on which there was water, but no soil or vegetation. A crate of breakfast cereal and a crate containing six living hens were also washed ashore. Which of the following strategies would enable the sailors to survive the longest?

A Feeding the breakfast cereal to the hens and then eating the hens.

B Feeding the breakfast cereal to the hens, eating the eggs that the hens produced, and then eating the hens when they died of starvation.

C Eating the hens and then eating the breakfast cereal.

D Eating the breakfast cereal, giving none to the hens, and then eating the hens when they died of starvation.

4 Many factors influence the size of populations. Choose the most accurate statement from the following.

A The size of a population that can be supported in an ecosystem is described as the supporting capacity of that ecosystem.

B Density-dependent factors, such as disease, affect a population size.

C The effect of density-independent factors, such as rainfall, fluctuate with population growth.

D The emigration of a large number of individuals causes density-independent factors to reach equilibrium.

5 When a dirt road falls into disuse, annual weeds soon begin to grow on it, increasing numbers. Eventually these weeds are no longer evident, as perennial grasses overgrow the area. The biological principle best used to explain this situation is:

A energy pyramids.

B ecological succession.

C cycling of matter.

D ecological equilibrium.

6 The process of crossing over and recombination

A produces identical daughter cells.

B occurs between homologous chromosomes.

C is different in oogenesis and spermatogenesis.

D increases mutation rates.

7 When a gene is read, single stranded mRNA is used. Initially pre-mRNA is formed, then mRNA is formed from the pre-mRNA. mRNA is different to pre-mRNA in that

A mRNA has both exons and introns removed.

B mRNA has non-coding exons removed.

C mRNA has coding introns removed.

D mRNA has non-coding introns removed.

8 Nucleosomes

A consist of DNA and housekeeping genes.

B contain genes that are continually expressed.

C contain a group of histone proteins with supercoiled DNA wrapped around them.

D are groups of histone proteins with methyl and acetyl groups attached.

9 Rett syndrome is one of the few known X-linked dominant disorders. It is a neurodevelopmental syndrome with symptoms appearing after 17 months of age. If a man with Rett syndrome marries a woman who does not have the disorder, which of the following is true for their children?

A All of their daughters and none of their sons will have Rett syndrome.

B All of their sons and none of their daughters will have Rett syndrome.

C Half of their daughters and half of their sons will have Rett syndrome.

D All of their children will have Rett syndrome.

10 The Isthmus of Panama was created 3 million years ago. It is a tiny strip of land that now joins North and South America. Before the isthmus existed, the ocean was home to 15 species of snapping shrimp. Today, there are 15 species of shrimp on one side of the isthmus and 15 different species on the other side. The difference in species on either side of the isthmus was probably caused by:

A allopatric speciation.

B sympatric speciation.

C post-reproductive isolation.

D macroevolution.

SHORT ANSWER QUESTIONS

1 Whales, goats, hippopotamuses, pigs and camels are mammals that share a common ancestor. There are two ideas about their evolutionary relationships.

The first idea (1) is shown on the left and the second idea (2) is shown on the right.

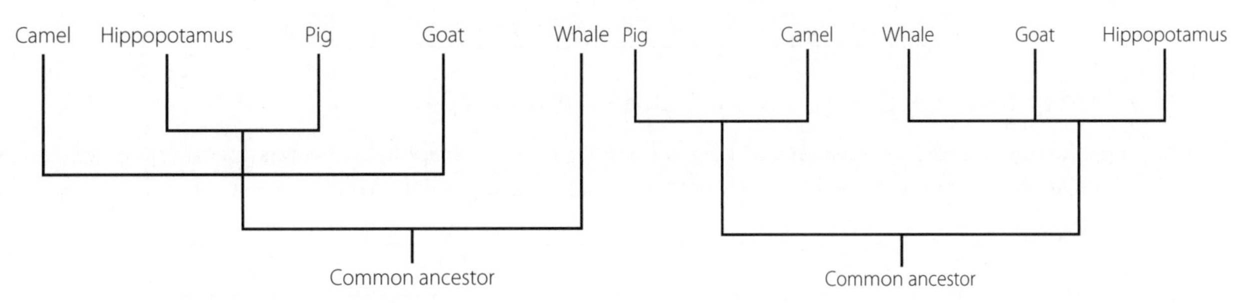

a Using information from the diagrams above, fill in the table to show which mammals share the most features and those with the fewest features in common with the pig.

	IDEA (1)	IDEA (2)
Mammals sharing the most features in common with the pig		
Mammals sharing the least features in common with the pig		

Whales are closely related to goats although they have many characteristics in common with sharks.

b Name the term used to describe the similarities in structure between whales and sharks.

____

c A student argued that whales and sharks do not have a common ancestor so they cannot be closely related. Explain if you agree or disagree with this idea and why.

____

____

d Suggest how studying molecular sequences could be used to determine which view of evolutionary relationships of these mammals is more accurate.

____

____

2 Biomass of the Daintree rainforest can be represented in a pyramid as shown below.

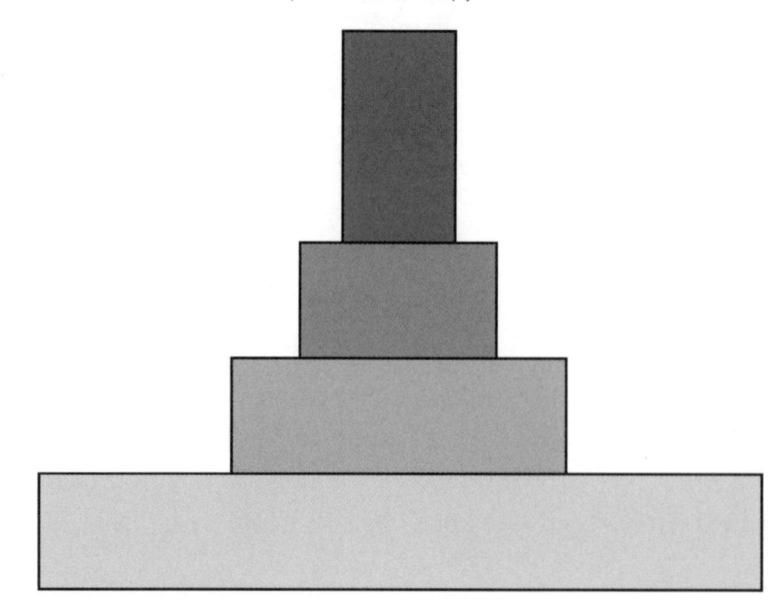

a Add labels on the pyramid of biomass to indicate the trophic levels.

b A useful rule in ecology is that about 10% of the energy at one trophic level is passed on to the next level. Suggest two reasons for the drop in biomass of organisms in passing from the first level to the fourth level.

Reason one: ______________________________

Reason two: ______________________________

The keystone species in the Daintree rainforest is the Southern Cassowary *Casuarius casuarius*. These birds mainly eat fruit but they can consume everything from small insects to mammals. After eating and digesting fruit, seeds are expelled in faeces. This allows for effective seed distribution.

c Explain why the Southern Cassowary is considered to be a keystone species.

____

____

d The Southern Cassowary interacts with other organisms through both symbiosis and predation. Provide evidence of these relationships.

____

____

e The Southern Cassowary is an endangered species. Predict the effect on the pyramid of biomass if the Southern Cassowary became extinct.

____

____

f Calculate the rate of growth and change in a Southern Cassowary population of 80 individuals if 20 new chicks hatched and survived, a mating pair entered the population and four males died in the following year. Show your calculations.

3 Achondroplasia is a form of dwarfism in humans and is inherited as an autosomal dominant condition. The gene for Achondroplasia is on chromosome 4 and most cases are the result of a new mutation. The mutation occurs as a single base substitution at nucleotide 1138 of the DNA sequence. The specific sequence of DNA in a normal allele is …..AGCTACGGGTGG….. and in a mutant Achondroplasia allele is …..AGCTACAGGTGG…..

a **…..AGCTACGGGTGG…..**

i Identify nucleotide 1138 by underlining or circling it and state the base change that occurs.

ii Write out the DNA complementary base sequence of the normal allele.

b Explain why the type of mutation causes one amino acid to change without affecting other amino acids in the protein.

c Predict what would happen to the protein if a nucleotide was inserted in position 1139.

d Two parents both with Achondroplasia can produce a child who is of normal height. Explain how this is possible by showing your working out of predictions of genotypes and phenotypes.

e Draw a labelled diagram of oogenesis. Use only chromosome 4 in the diagram making sure alleles are labelled.

f Crossing over is a normal part of oogenesis. Explain the biological significance of crossing over.

g DNA replicates during oogenesis. Use words or fully labelled diagrams to describe the steps involved in DNA replication.

h Explain how the products of some genes regulate the products of other genes.

4 Stick insects are commonly raised and maintained in classrooms. There are around 200 species naturally occurring in Australia. Stick insects are found in a range of habitats. The Lord Howe Island stick insect, *Dryococelus australis*, was driven to the brink of extinction by Black Rats in the early twentieth century. However, in 2001 it was rediscovered on Balls Pyramid, a rat-free volcanic outcrop 23 km off the coast of Lord Howe Island. Melbourne Zoo staff are undertaking a captive breeding program for this species.

a Explain the effect of constant environmental conditions on the diversity of Lord Howe Island stick insects bred at the zoo.

b Outline the steps involved in the allopatric speciation of Lord Howe Island stick insects.

9780170411745

c Explain how genetic drift could have been a contributing factor.

d Grasshoppers and stick insects are sometimes confused with each other as they look similar. Name and describe the pattern of evolution that gave rise to these similar but not closely related species.

e Many female phasmids do not always need to mate in order to produce offspring. This form of reproduction is called parthenogenesis and all the eggs produced will hatch into females. Suggest how this influenced the drastic reduction in numbers with the introduction of Black Rats early in the twentieth century.

ANSWERS

CHAPTER 1 REVISION

1.1 THE GREAT BARRIER REEF: BIODIVERSITY HOTSPOT

1. Biodiversity is the full range of living things in a particular area or region. The Great Barrier Reef has more than 1500 species of fish, 400 species of coral, 4000 species of mollusk and 240 species of birds, along with a great diversity of other organisms.
2. The Great Barrier Reef contains 2500 individual reefs within its range. It includes extensive cross-shelf diversity from the low water mark up to 250 km offshore and ranges from shallow inshore areas to waters more than 2 km deep. This variety of habitat shelters thousands of species.
3. Two of the most important species that the Great Barrier Reef houses are the dugong and the large green turtle, it contains an enormous variety of oceanic habitats and their accompanying species and many of the small islands are home to globally important breeding colonies of seabirds and marine turtles.
4. Student responses will vary widely but should consider whether they believe World Heritage status is required to protect important areas and whether these areas are sufficiently protected by this status.

1.2 THE FIVE MEASURES OF SPECIES DIVERSITY

1

MEASURE OF SPECIES DIVERSITY	DEFINITION	EXAMPLE
SPECIES RICHNESS	*The number of species present in an ecosystem*	*eg. 3 different species*
RELATIVE SPECIES ABUNDANCE	*The number of individuals present for each species in an ecosystem*	*eg. 2 Eucalypt sp., 6 Melaleuca sp. and 3 Acacia sp.*
PERCENTAGE COVER	*The percentage of the quadrat that a species takes up*	*eg. Eucalypt sp. covers 40% of a quadrat*
PERCENTAGE FREQUENCY	*The percentage of quadrats that a species appears in*	*eg. Eucalypt sp. appears in 3% of quadrats taken*
SIMPSON'S DIVERSITY INDEX (D)	*The combined ratio of individuals in each species to the total individuals in an ecosystem*	$D=1-\left(\frac{\sum n(n-1)}{N(N-1)}\right)$ $D=1-\left(\frac{2\times1+6\times5+3\times2}{11\times10}\right)$ $D=0.655$

1.3 SIMPSON'S DIVERSITY INDEX

1. Site 1: 1-(236/870) = 0.729

   Site 2: 1-(208/870) = 0.761

   Site 3: 1-(920/1722) = 0.466

   Site 4: 1-(56/306) = 0.817
2. Least biodiverse is Site 3, then Site 1, Site 2 and Site 4. This should have been predicted from the over-abundance of *Bouteloua* in Sites 3 and 1 and the evenness of distribution in Site 4.
3. Site 4 is more biodiverse because each species is balanced in its occupation of the area. Site 3 may have more life, but the over-abundance of *Bouteloua* at the expense of *Banksia* and *Callistemon* species reduce its biodiversity considerably.
4. D for Site 2 is 0.761 and D for Site 4 is 0.817. Both are quite diverse; however, the difference between the highest and lowest number of individuals in Site 2 is 7, while in Site 4 is only 4. The smaller range in Site 4 contribute to its higher D.
5. Site 1: 1-(104/306) = 0.660 The loss of *Bouteloua* has reduced the biodiversity of this site.

   Site 2: 1-(136/420) = 0.676 The loss of *Bouteloua* has also reduced the biodiversity of this site.

   Site 3: 1-(50/132) = 0.621 The loss of the over-abundant *Bouteloua* has improved the biodiversity of this site to just less than Sites 1 and 2.

   Site 4: 1-(26/132) = 0.803 The loss of *Bouteloua* has reduced the biodiversity at this site, but not by very much, given its balance with the other species.

1.4 SPATIAL AND TEMPORAL COMPARISONS

1

TEMPORAL SCALE	AT DIFFERENT TIMES OF DAY	AT DIFFERENT TIMES OF YEAR	AT DIFFERENT TIMES OVER A DECADE
How species diversity may differ in the same place:	eg. many more species active during the day than at night	eg. many more species of insect active during the summer months than over winter	eg. extended periods of drought may cause species to move away for years at a time
SPATIAL SCALE	**IN DIFFERENT LOCAL AREAS**	**IN DIFFERENT STATES**	**ON DIFFERENT CONTINENTS**
How species diversity may differ at the same time:	eg. more species active in a forest than in a city	eg. more species active in North Queensland in winter than in Victoria	eg. more species active in Australia than Antarctica

1.5 ENVIRONMENTAL LIMITING FACTORS

1

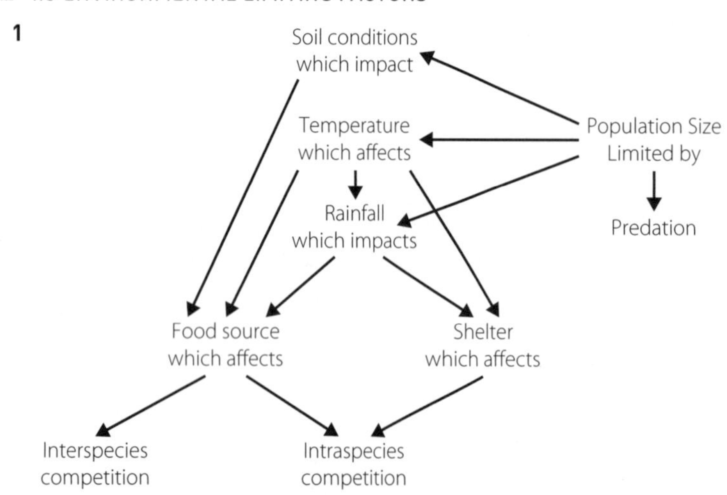

CHAPTER 1 EVALUATION

1 C

2 A

3 Biotic factors include: predation, competition and food sources. Abiotic factors include: rainfall, temperature and soil conditions.

4

a Species richness, species abundance, percentage cover, percentage frequency and Simpson's diversity index.

b Species richness, percentage cover and percentage frequency.

c Species richness, species abundance and Simpson's diversity index.

5 Student responses will vary but should consider the number of species using an urban street during the day versus at night and the number of individuals of each species present.

6 Analyse:

	LOCAL WOODLAND SITE	BACKYARD SITE
SPECIES RICHNESS	6	5
SPECIES ABUNDANCE	As in table. Relatively even distribution	As in table. Very large differences present
SIMPSON'S DIVERSITY INDEX	$D=1-\left(\frac{\sum n(n-1)}{N(N-1)}\right)$ $D=1-\left(\frac{4\times3+5\times4+1\times0+6\times5+1\times0+2\times1}{19\times18}\right)$ $D=0.187$	$D=1-\left(\frac{\sum n(n-1)}{N(N-1)}\right)$ $D=1-\left(\frac{6\times5+3\times2+10\times9+1\times0+2\times1}{22\times21}\right)$ $D=0.277$

Communicate:

Despite the species richness and species abundance appearing to support the student, that the local woodland site has greater biodiversity, Simpson's diversity index does not support the student. The diversity index takes into account the fact that two of the woodland's six species have only one individual present, which is not appropriate to support strong biodiversity. The abundance of *Westringia* sp. in the backyard site appears to have less of an impact on the final diversity index than the lack of individuals at the woodland site.

CHAPTER 2 REVISION

2.1 CLASSIFICATION AS A DIAGNOSTIC TOOL

1 Doctors need to be able to identify species of bacteria because some strains of bacteria cause similar symptoms but require very different treatment.

2 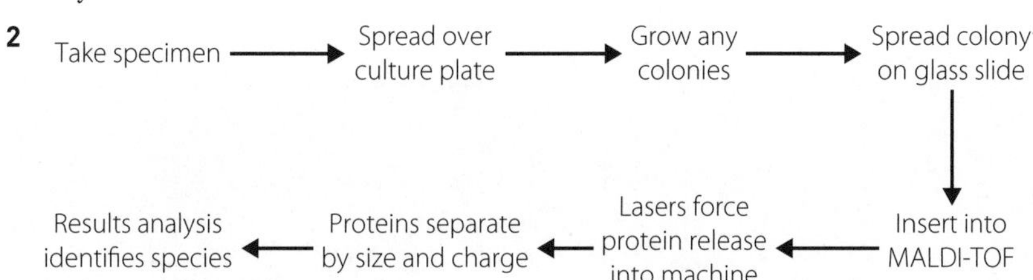

3 Traditional identification uses the gram status (positive or negative) and shape of a bacteria, along with other chemical tests, to identify species, while MALDI-TOF mass spectrometry uses the sizes and charges of the range of proteins the bacteria produces to identify species.

4 Engineers and physicists created the MALDI-TOF technology to enhance the efficiency of a process in microbiology, which directly affects the medical community.

2.2 LINNAEAN CLASSIFICATION

1

LEVELS OF CLASSIFICATION	EXAMPLES
Kingdom	Animalia (animals), Plantae (plants) etc.
Phylum	Chordata
Class	Mammalia
Order	Carnivora
Family	Felidae
Genus	*Felis*
Species	*Felis catus*

2 Student responses will vary but two examples include: the Red Panda (*Ailurus fulgens*), which was originally classified in Family Ursidae and Family Procyonidae before being reclassified into its own family, Family Ailuridae; and Manatees, which were thought to be related to walruses (*Odobenus rosmarus*; Order Carnivora; Family Odobenidae), but have since been reclassified into Family Trichechidae, with the dugong as the only other member of its order, Sirenia.

2.3 K/R SELECTION

1

K SELECTION	R SELECTION
Whales	*Mice*
Elephants	*Plants*
Humans	*Dogs*
Magpies	*Ants*
	Locusts
	Crocodiles
	Turtles
	Salmon
	Coral

2 Some organisms, such as crocodiles and dogs, have fewer offspring and more parental effort than, for example, coral and turtles. However, they still have significantly more than two offspring each year and only parent them for a few weeks or months, rather than the year or more that the other K selection organisms do.

3 Student responses will vary but two acceptable answers could be to include a third, moderate selection that would encompass the organisms that have between 2 and 12 offspring each year, or to have a single boundary between K and r organisms, perhaps more or less than 6 offspring.

2.4 CLADISTICS

1

	STARFISH	SNAIL	BEE	COW	DUCK	TURTLE
BILATERAL SYMMETRY	*a*	*P*	*P*	*P*	*P*	*P*
HAS A BACKBONE	*a*	*a*	*a*	*P*	*P*	*P*
GIVES BIRTH TO LIVE YOUNG	*a*	*a*	*a*	*P*	*a*	*a*
EG. WARM-BLOODED	*a*	*a*	*a*	*P*	*P*	*a*
EG. HAS LEGS SEPARATE FROM THE BODY	*a*	*a*	*P*	*P*	*P*	*P*

2-5

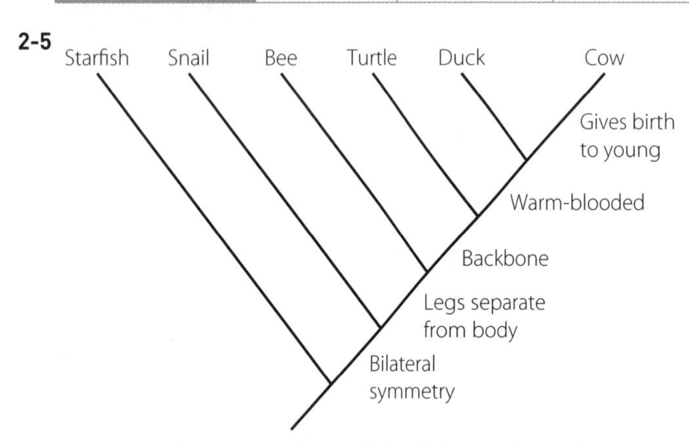

6 Any group of organisms that include all descendants of a common ancestor.

7 Any reasonable, logical explanation, including choice of traits.

8 A cladogram represents the simplest and most logical evolutionary relationships between the selected organisms.

2.5 UNDERLYING ASSUMPTIONS OF CLADISTICS

1

ASSUMPTION	DESCRIPTION	EXAMPLE
Common ancestry	As all life evolved from a single ancestor, any group of organisms will share a common ancestor at some point in the past.	The first life evolved into all life, therefore the rabbit and the whale must share some evolutionary relationship.
Dichotomous divergence	The offspring of an ancestral species diverge dichotomously in a process called cladogenesis.	Offspring will either be very similar to their parent or different in one way. The offspring that are different in one way may form a new species.
Increasing difference over time	Organisms become increasingly different as they continue evolving from their point of cladogenesis.	Organisms that are closely related, such as chimpanzees and humans, share many similarities, while organisms that are more distantly related, such as humans and snails, share comparatively fewer similarities.

2.6 DETERMINING EVOLUTIONARY RELATEDNESS

1

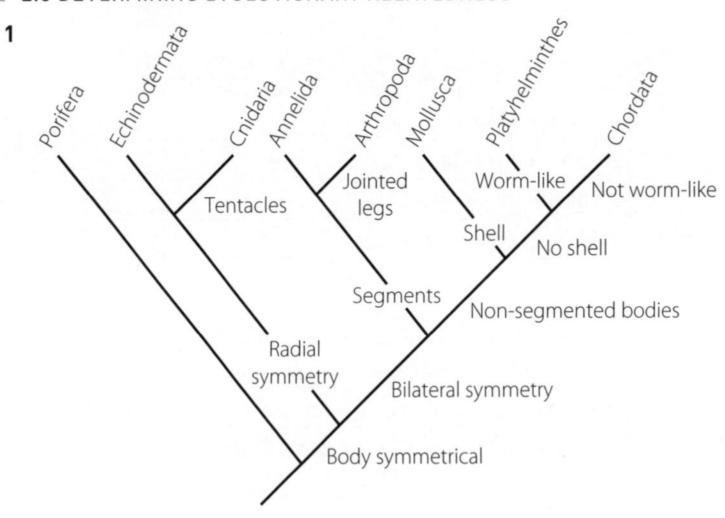

2

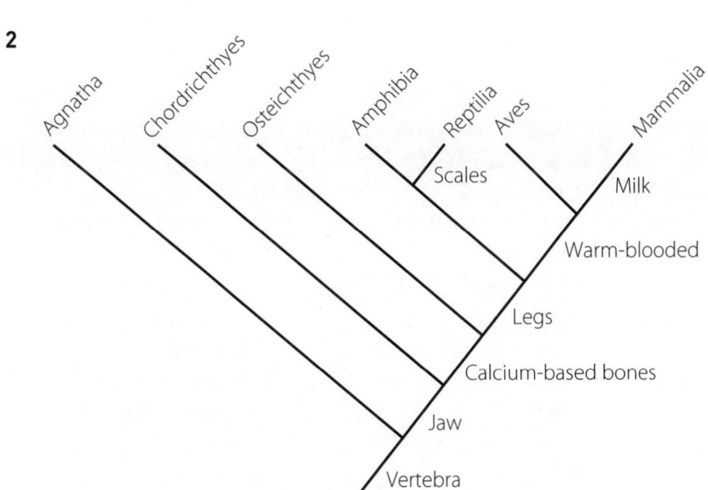

2.7 MULTIPLE DEFINITIONS OF SPECIES

1 A biological species is a group of interbreeding, fertile organisms. A morphological species is a group of organisms with very similar physical features. A phylogenetic species is the smallest group of organisms who can all trace their origins to a single ancestor.

2 Biological species concept: Separate species because although they can interbreed their offspring are not fertile.

Morphological species concept: Similar species because they have very similar physical features.

Phylogenetic species concept: Separate species because donkeys and horses share a common ancestor with zebras.

3 Limitations: the public can be confused when new species are discovered or redefined based on different concepts and each species requires its own scientific name, which is difficult to determine when three different concepts can be applied.

Benefit: organisms can be grouped with other organisms using different concepts, which enhances the flexibility of scientific study.

2.8 HYBRIDS

1 Student responses will vary widely but should include a realistic hybrid, the features it brings from its parents and the two names used for the offspring.

2.9 SPECIES INTERACTIONS

1

COMPETITION	PREDATION
An owl stealing another owl's prey	A snake eating a mouse
A male kangaroo fighting another for females	A ramora fish eating shark parasites
A magpie swooping a cyclist	A cane toad eating a frog
Two colonies of ants fighting over a dead grasshopper	A feral cat eating a parakeet
	A possum eating a eucalypt flower
SYMBIOSIS	**DISEASE**
A clownfish living within a sea anemone	A rabbit infected with the myxoma virus
A tapeworm colony in the gut of a dog	A koala infected with chlamydia
A cattle egret riding on a cow to eat the insects disturbed by its feeding	A Tasmanian devil with a Devil Facial Tumour
An epiphyte growing in the canopy of a tall tree	

2.10 CLASSIFYING ECOSYSTEMS

1

CLASSIFICATION SYSTEM	FEATURES	BENEFITS	DRAWBACKS
HOLDRIDGE LIFE ZONE	Uses temperature, evapotranspiration and precipitation to determine zone.	Quite simple. Requires only three broad data points. Don't have to see the area.	Lacks finesse. Areas could easily fall into more than one zone.
SPECHT'S	Uses tallest vegetation and its foliage cover to determine classification.	Very simple. Only two estimated data points required. Fits most Australian terrestrial ecosystems.	Must have a visual of the area. No allowance for temperature to affect classification.
ANAE	Aquatic ecosystem classification system using more than 20 different metrics to determine classification.	Fits all aquatic ecosystems, both fresh and saltwater. High level of detail and specificity.	Very complex. Requires extreme detail to classify an area.
EUNIS	European-based habitat system that covers terrestrial and aquatic systems, both natural and artificial.	Comprehensive coverage of habitats.	Very complex. Different metrics used depending on classification. European-based.

2 Student responses will vary but should consider both the variety of ecosystems in Australia and the consistency of a single, universal system.

2.11 STRATIFIED SAMPLING

1 Stratified sampling is a statistical sampling technique that divides an area into strata for separate sampling.

2 In ecology, gathering data about an entire population is difficult and statistical sampling techniques must be employed to form accurate estimations. Stratified sampling can be used to accurately estimate population size, density and distribution over an area, as well as identify and define environmental gradients and habitat zones.

3 Ideal sites for stratified sampling are ecosystems or ecoregions that are comprised of a wide variety of subsections and habitats. When the abiotic and biotic factors vary widely across an ecosystem, stratified sampling can ensure the data reflects a more accurate portrayal of the area.When selecting strata within the area, it is important to consider how large an area is appropriate. The boundaries between strata should be as defined as possible.

4 A quadrat is a square, the size of which is determined according to the organism being studied, measured at ground level. It is most often used when measuring plant or fungus species because they are stationary. For each quadrat, the number of individuals of each species is counted (or estimated in the case of difficulty) and recorded. The data from multiple quadrats is then averaged. A transect is a line drawn through a community to provide a boundary for sampling. Again, this is a useful method when species are fixed in place, such as plants and fungi. In order to improve the data collected, quadrats may also be placed at intervals along the transect line.

9780170411745

5 To minimise bias in the data, sufficient quadrats should be taken in each stratum to ensure the sample is representative. However, the number of quadrats from each stratum should be proportional to the relative size of each habitat. Other processes for minimising bias in stratified samples include using a random number generator to determine the position of quadrats taken within a stratum and following strict criteria for what constitutes an individual in the case of groundcover plants where individuality is difficult to determine and how much of the plant must be within the quadrat to count it. Any equipment used for abiotic sampling should be calibrated immediately before use and the associated precision of the instrument should be noted in the data.

6 The simplest presentation of sampling data is in table form, though it is easier to spot patterns and trends when data is in a visual format, such as a graph. Communication of ecological data is usually through publishing investigation papers in scientific journals.

CHAPTER 2 EVALUATION

1 B

2 C

3 Domain, Kingdom, Phylum, Class, Order, Family, Genus and Species.

4 a *Clostridium difficile* and *Escherichia coli.*

b *Clostridium difficile.*

c *Neisseria* spp. cannot be identified to species level. Several more questions are likely to be required to distinguish between these gram negative cocci species.

5 Student responses will vary but should be justified by prior knowledge of the morphological and reproductive similarities between the three groups.

6 Student responses will vary but arguments should be clear, logical and well-supported by evidence.

CHAPTER 3 REVISION

3.1 FOOD CHAINS

1

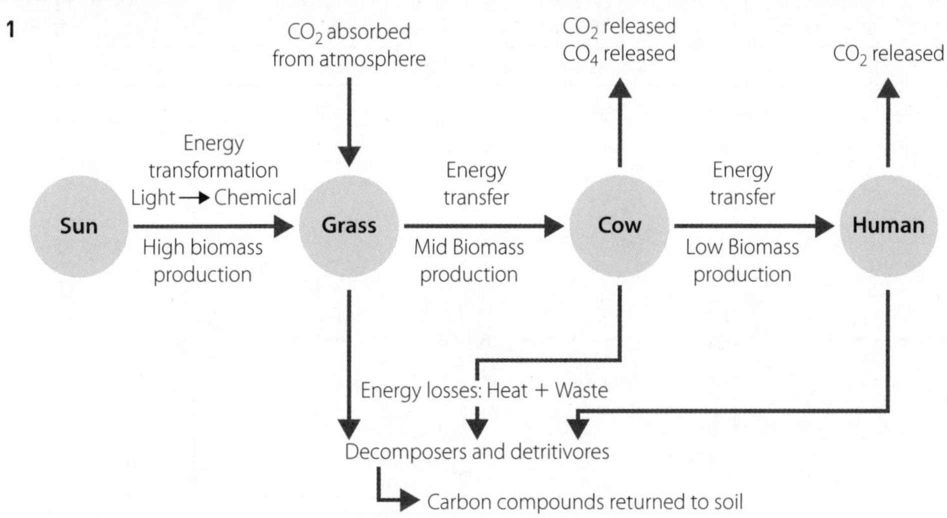

3.2 CALCULATING ENERGY TRANSFERS

1 a 5 kJ

b 0.1 kJ

c 30 kJ

2 a Grass 1%, Rabbit 40%, Snake 5%.

b Eucalypt 0.06%, Koala 16.7%.

c Tree 10%, Caterpillar 20%, Bird 8.3%, Cat 18%.

3 a

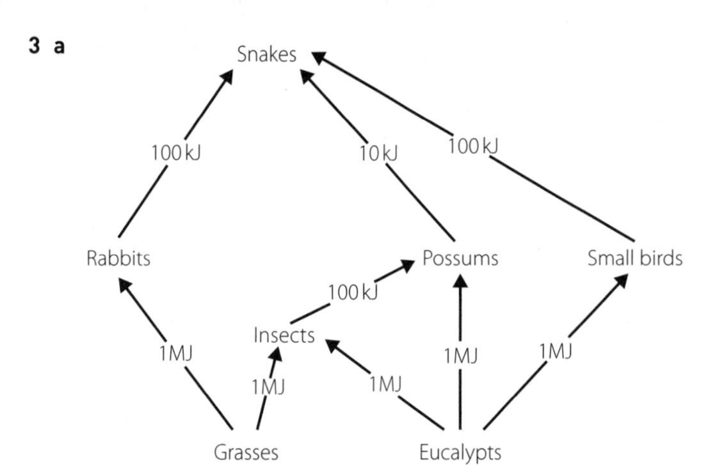

b If dingos were added to this ecosystem they may eat snakes (10 kJ/day), possums (10 kJ/day) or rabbits (100 kJ/day). On paper, this would support a total of 24 dingoes, except that the snakes would be reduced to eating only small birds, which would affect their population. A small pack of dingoes (around 10) may reasonably be able to live here but they would quickly outgrow the food sources.

3.3 ENERGY FLOW DIAGRAMS

1 a

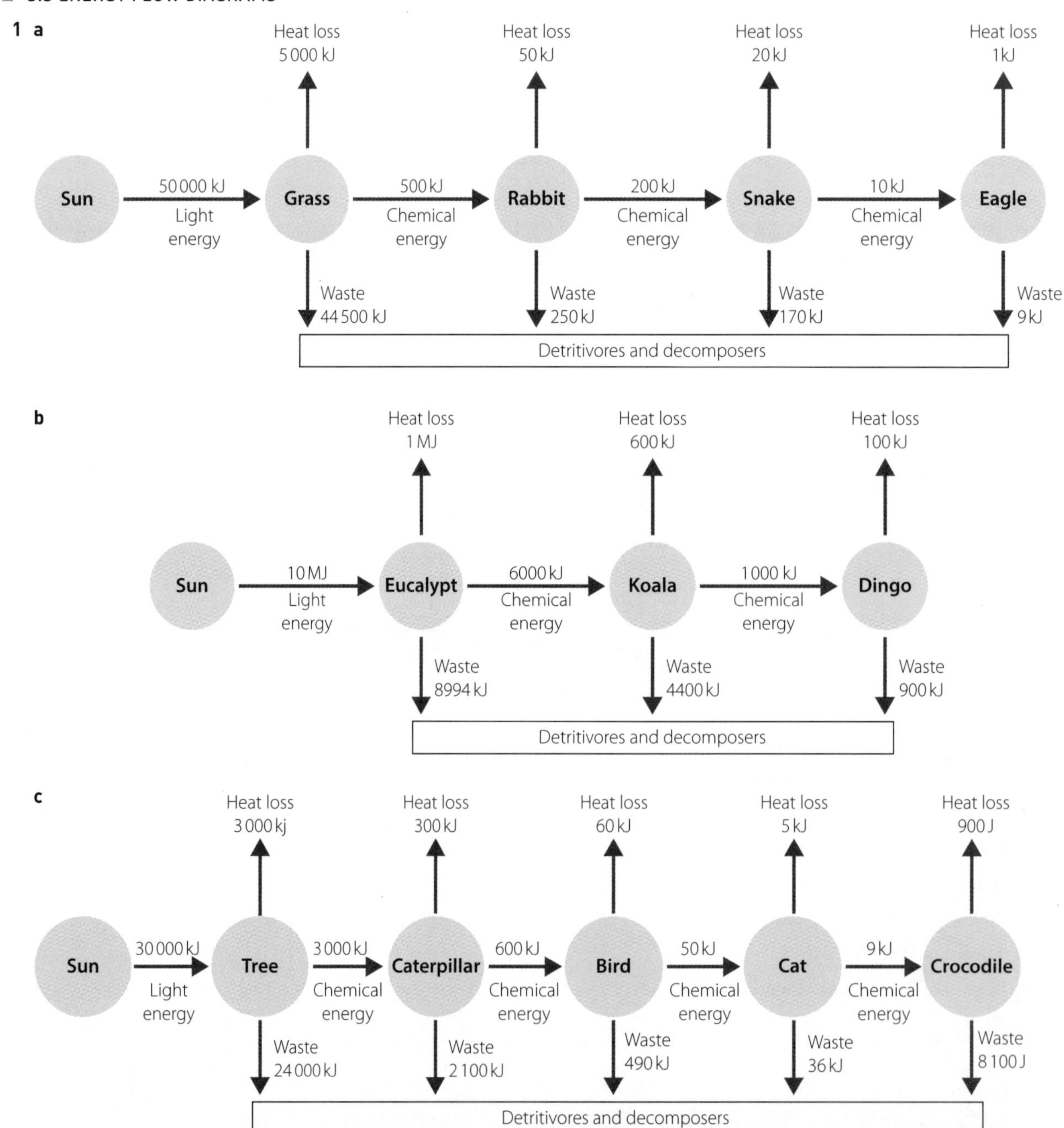

2 Two limitations of energy flow diagrams include the fact that the sun's energy is actually dispersed in a web rather than a chain, so not all areas of the journey are mapped in the diagram, and the simplified format means that the recycling of energy from the detritivores and decomposers is not included.

3.4 THE CARBON CYCLE

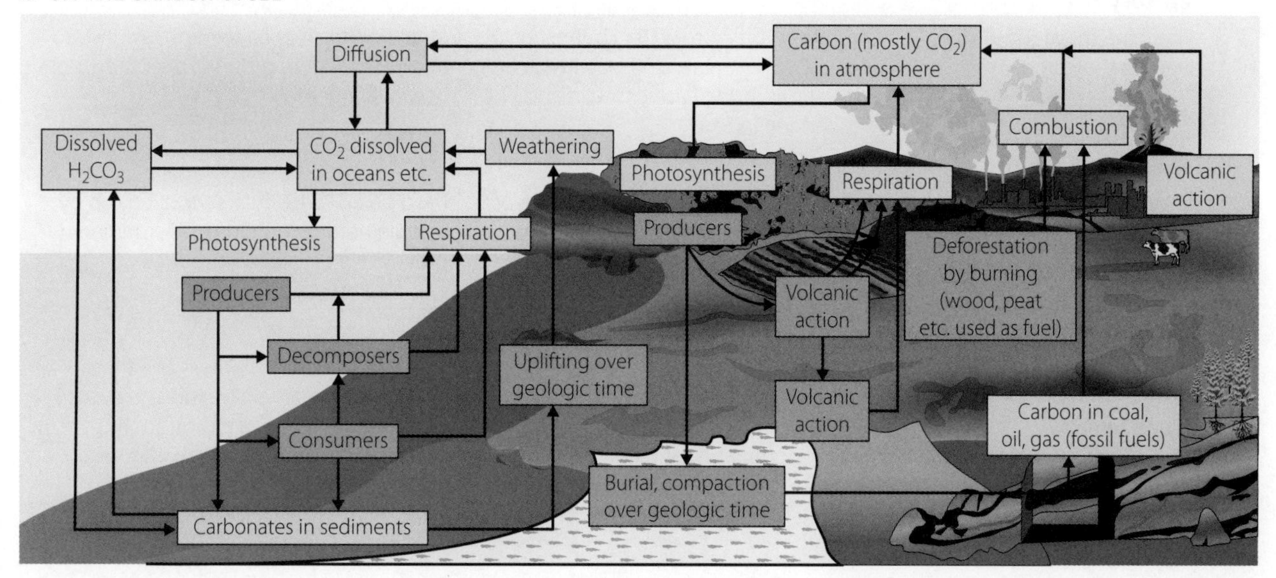

3.5 THE NITROGEN CYCLE

1 Major differences between the nitrogen and carbon cycles include: the nitrogen cycle is composed of two cycles, elemental and ionic, while the carbon cycle is a single cycle of compounded carbon; the nitrogen cycle requires organisms (nitrogen fixers and denitrifiers) to interface between the two cycles, while the carbon cycle does not require the input of organisms; and organisms can interact with atmospheric carbon directly through respiration and photosynthesis, while organisms (in general) cannot interact directly with atmospheric nitrogen.

2 A build-up of antibiotics in the soil and water would negatively affect the populations of bacteria that live in soil and water. Since many of the nitrogen fixers and denitrifiers are bacterial, these vital organisms would not be able to cycle the nitrogen effectively and there may be a build-up of nitrogen in the atmosphere and limits the availability of ionic nitrogen for plants and animals.

3.6 THE WATER CYCLE

1 An increase in global temperature will increase the rate of evaporation from the ocean, land and plants, which will form more clouds and drop more precipitation. This will make the cycle faster, causing more frequent rains and storms. Raising global temperatures will also cause ice sinks, such as glaciers and permafrost, to melt, which will increase the free-flowing water available in the water cycle. This may lead to permanent flooding of low-lying areas, raising the water table in groundwater stores and a further increase in rains and storms on land.

2 Examples include: melting ice floes, glaciers and permafrost; more frequent and violent rains and storms; more frequent and more permanent flooding; raising the water table; and diluting nutrients for plants and aquatic life.

3 An imbalanced water cycle is likely to rebalance eventually as it copes with a larger volume of free-flowing water on the planet by cycling faster and with more force. However, the rebalanced cycle is unlikely to be healthy for current terrestrial and aquatic life, which will need to adapt to the new stabilising conditions.

3.7 ECOLOGICAL NICHE

1

	GREAT WHITE SHARK	HUMPBACK WHALE
Territory Where are they found? How deep do they range?	Coastal and offshore waters between 12 and 24°C, as deep as 1200 m, with greater populations in the Pacific Ocean and Mediterranean Sea.	Migration from polar feeding grounds to tropical breeding grounds over the whole ocean. Range down to 200 m.
Feeding behaviour What do they eat? How do they eat it? Is there anything they won't eat?	Opportunistic feeding and scavaging on all forms of marine mammal and large fish. High-fat prey is preferred. Chunks twisted off by rotating and thrashing.	Feed on krill and small schooling fish by bubble netting, gulping and stunning prey with fluke slapping.
General activity When do they sleep? When do they eat? Do they fight/defend territory?	They don't sleep, but actively hunt at the surface within 2 hours of sunrise. They don't defend territory as such but will fight for dominance at feeding grounds.	Thought to sleep half-brain at a time, to facilitate breathe/dive/surface cycle. Generally diurnal and although they don't appear territorial, they may sing to announce their area.
Social/reproductive behaviour How many are found together? How and when do they reproduce? How many offspring do they have?	Limited social contact. Sexual maturity occurs by 26 (males) and 33 (females). An 11 month gestation period produces 2-10 pups every few years.	Not generally social animals. Individuals gather for migration and feeding but not for long. Sexual maturity around 5-7 years. A single calf gestates for 11.5 months every two or three years.

3.8 DATA ANALYSIS: THE COMPETITIVE EXCLUSION PRINCIPLE

1

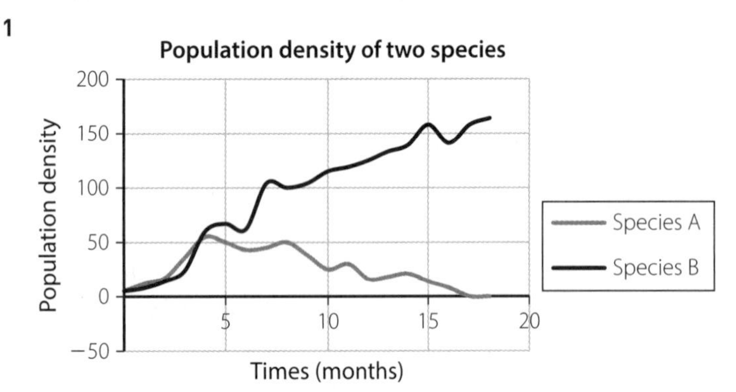

2 Species A

3 Approximately 4 months before competitive exclusion begins, approximately 16 months before Species A is eliminated.

4 Species B

5 Species A begin quite strongly but at 4 months, Species B began to compete with them. They held steady for a period of 4 months until they began dying off over the next 8 months.

6 This scenario shows that an ecosystem cannot support two species that occupy the same niche, so many niches must be present for many species to enhance the biodiversity of an area.

7 Gause's competitive exclusion principle appears to be supported by this experiment. His claim that two species occupying the same niche would be able to coexist for a short period of time is shown in the first 4 months of the experiment, when both Species A and B increase their population density. His further claim that they would compete with each other for space and resources is also apparent in the next 12 months, when Species B continues to grow and thrive while Species A dies off in the area.

8 If a disease affected the reproductive success of Species B at 10 months, their growth rate would drop considerably. Given that Species A made a small recovery around 10 months, it is possible that the stronger reproduction of Species A may have given it a competitive advantage and allowed it to reclaim some of the space and resources from Species B. If the disease was not resolved, Species B may have eventually been outcompeted by Species A.

9 Species C has a shorter gestation period and so, will produce offspring quicker. This gives it a competitive advantage in that any free spaces opening on the rock face will likely be settled by Species C. It is unlikely that it will outcompete either of Species A or B in the long run, given their headstart on the rock face.

10 Student responses will vary but should consider the means of measuring Species C, the specific parameters to be tested to measure 'impact' (changes to other populations, changes to abiotic conditions etc) and the length of time required to obtain reasonable data on the ecosystem's response to Species C.

3.9 KEYSTONE SPECIES

1

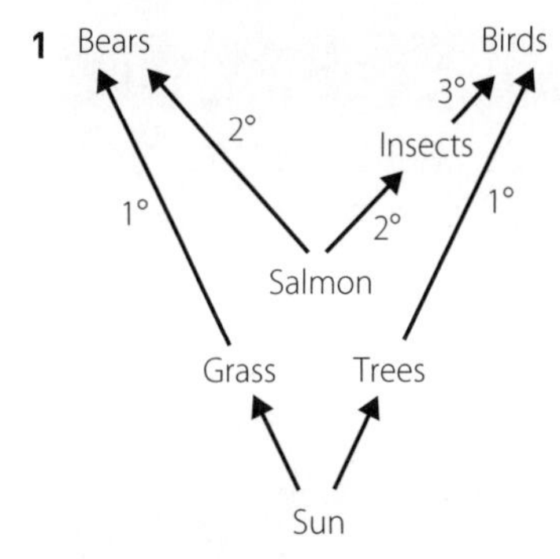

2 Migratory salmon

3 They eat copious quantities of salmon and discard the remains on the forest floor, which feeds the grasses that they eat in large amounts in spring.

4 Bioaccumulation

5

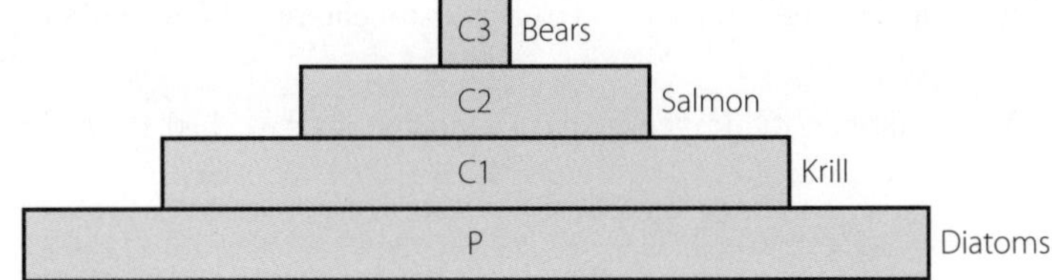

6 Without bears, the salmon would overpopulate the local waterways. Without the salmon remains littering the forest, there would be little grass growth in the spring, limited insect laying areas and thus, less food for the birds and other wildlife in the forest.

CHAPTER 3 EVALUATION

1 B

2 D

3 Energy flow diagrams include the value and type of energy flowing between the trophic levels, whereas food chains only denote that energy flow of some sort exists between organisms.

4 20 kJ

5

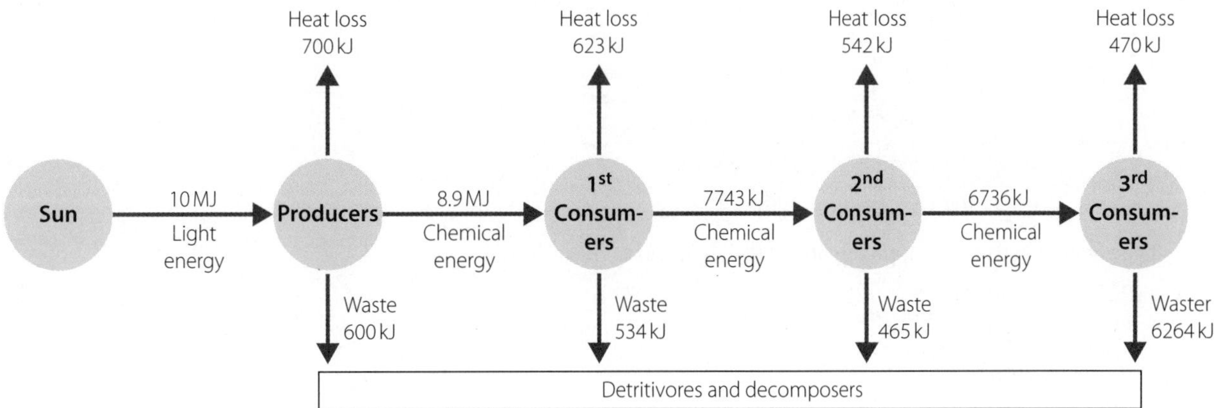

6 Ecological niche is the role and space that an organism fills in an ecosystem, including all its interactions with the biotic and abiotic factors of its environment.

7 Student responses will vary widely but arguments should be logically consistent, concisely stated and clearly supported by evidence.

8 Student responses will vary but should consider the limited dispersal of tree seeds without Cassowaries, the impact this would have on interspecies competition and the strong role of rainforest trees as food and shelter for other species.

CHAPTER 4 REVISION

4.1 CARRYING CAPACITY OF POPULATIONS

1 The carrying capacity describes the maximum population size of a species that can be supported in a given environment.

2

BIOTIC LIMITING FACTORS	ABIOTIC LIMITING FACTORS
Food supply	Acidity
Competition	Rainfall
Disease	Humidity
Parasites	Temperature
Predation	Salinity
Mates	Shelter
	Light

3 Catastrophic events: Volcanic eruption, tsunami, fire, drought, earthquake, flood, cyclone.

4 There is exponential growth between years one and three. Food and resources are plentiful.

5 Carrying capacity is around 70. An equilibrium is reached at around 70.

6 Student answers will vary but can include: food supply could be reduced because of competition between kangaroos or an introduced species; a disease could have killed some of the kangaroos; there could be a predator entering the area; abiotic factors could have changed.

7 Carrying capacity decreased in the last four years. As the numbers didn't recover and an equilibrium is reached, it is likely a long term factor decreased the carrying capacity at this time.

4.2 POPULATION CHANGES

1 a $\text{PGR} = (\text{br} + \text{ir}) - (\text{dr} + \text{er})$

$= (11 + 5) - (4 + 7)$

$= 16 - 11$

$= 5$

$= \frac{5}{100} \times \frac{100}{1}$

$= 5\%$

The population will grow by 5 individuals per 100 every year, or by 5%.

b

TIME	POPULATION SIZE
T_0	135
T_1	135 + 5% of 135 = 141.75 = 141
T_2	141 + 5% of 141 = 148.05 = 148
T_3	148 + 5% of 148 = 155.40 = 155
T_4	155 + 5% of 155 = 162.75 = 162
T_5	162 + 5% of 162 = 170.10 = 170

The population would be approximately 170 individuals in the fifth year.

2 a $\text{PGR} = (\text{br} + \text{ir}) - (\text{dr} + \text{er})$

$= (93 + 22) - (36 + 42)$

$= 115 - 78$

$= 37$

$= \frac{37}{1200} \times \frac{100}{1}$

$= 3.08\%$

The population growth rate is 37 individuals per 1200 every year or 3.08%.

9780170411745

b

TIME	POPULATION SIZE
T_0	1564
T_1	1612
T_2	1661
T_3	1712

The population would be approximately 1712 individuals in the third year.

c Students' answers will vary, and could include any of the following suggestions: disease, over-fishing, lack of food resources, unseasonal change in climate, increase in predators.

4.3 MEASURING POPULATIONS

1 Direct observation is time consuming and not accurate when used for mobile populations that move in and out of areas.

2 Two of: using satellite images, aircraft, drones or other reasonable suggestions.

3 Individual animals are captured, marked and released. After a time, the population is re-sampled and the number of marked animals caught gives an indication of population size.

4 a 16 years ago N = 510; Currently N = 153

b The population size of the turtles has decreased by about one third after sixteen years.

c There is no population change through migration, births or deaths between the sampling periods. All animals are equally able to be caught (individuals are not 'trap happy' or 'trap shy' meaning the same individuals will be sampled over and over). Marked animals are not hampered in their ability to move and mix freely with the rest of the population.

5

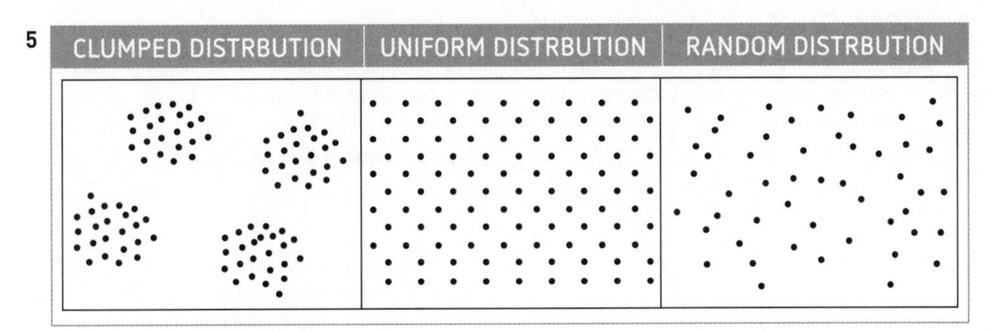

4.4 POPULATION GROWTH PATTERNS

1 a

TIME	NUMBER OF BACTERIUM
0	2
2	4
4	8
6	16
8	32
10	64
12	128
14	256
16	512
18	1024
20	2048
22	4096
24	8192
26	16 384
28	32 768
30	65 536
32	131 072

b

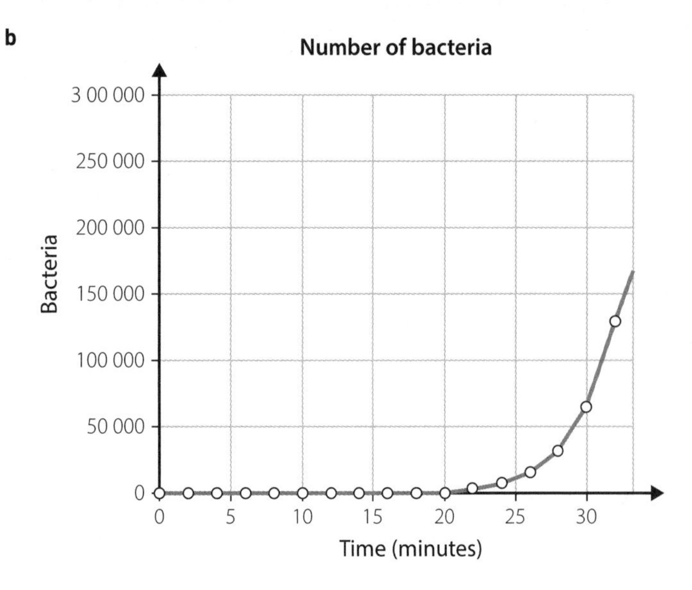

c Exponential growth J-curve.

d The population growth steadies because the nutrients available to the bacteria limits the population's growth rate.

e Logistic growth S-curve.

2 Exponential growth is possible under ideal conditions with unlimited resources. The world's human population is a good example.

3

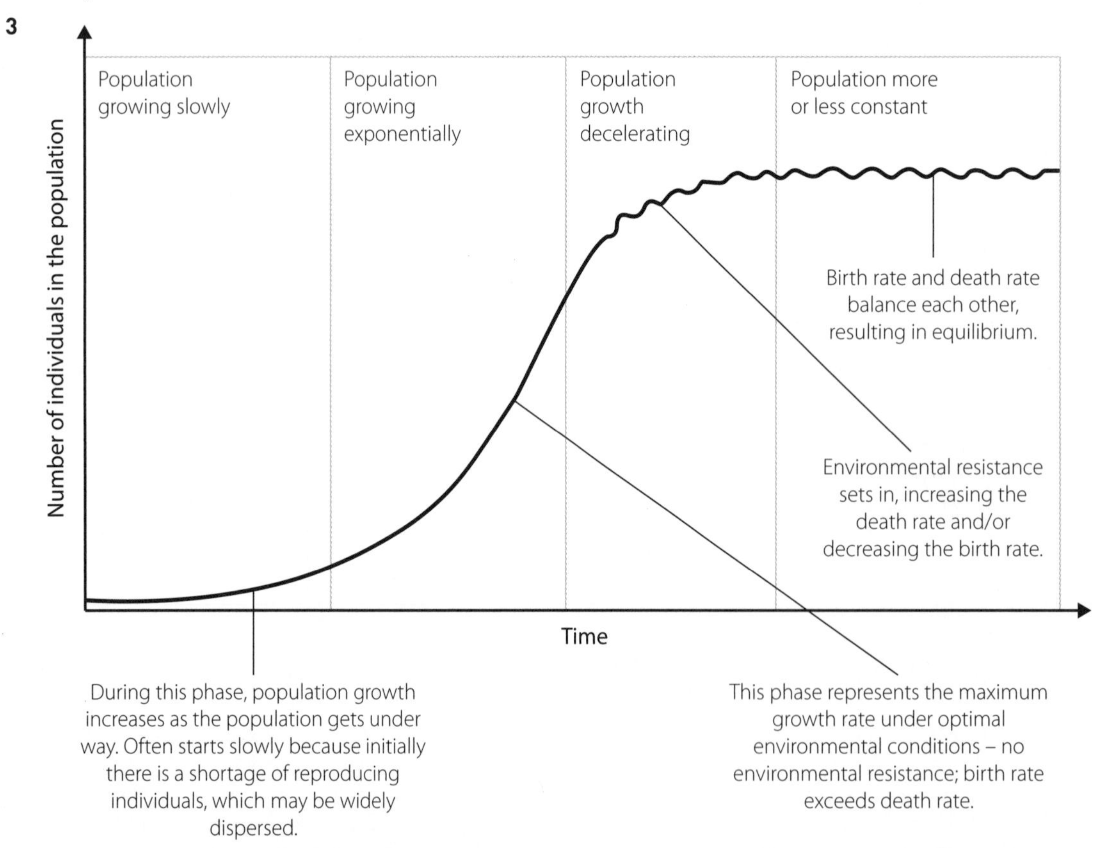

CHAPTER 4 EVALUATION

1 The area cleared for logging would show exponential growth. There would be more resources for opportunistic species that move in and colonise the area as quickly as they are able.

2 Resources will be used up and competition will limit population growth.

3 a The population is increasing by 20 rainbow lorikeets per year.

b Immigration and emigration.

4 Student answers will vary but can include

Biotic factors: Food supply, competition, disease, parasites, predation, availability of mates.

Abiotic factors: Rainfall, humidity, temperature, shelter, light.

5 **a** The sampling technique is called the Lincoln Index method and the formula is $N = \frac{M \times n}{m}$

b $N = \frac{200 \times 350}{70}$

$= 1000$

CHAPTER 5 REVISION

5.1 ECOLOGICAL SUCCESSION

1 Bare rock left after volcanic eruption → In time, mosses and lichens start to colonise the rock. As these die, organic matter is added to weathered rock particles making simple soils → As the soils develop, grasses and small herbaceous plants start to grow. More organic matter is added and roots of plants aid break up of rock material → Deeper soils hold more water, small shrubs colonise these better soils. Nutrient availability increases; more root action → Eventually trees establish, leading to the development of a climax community on mature soils.

2 Type of succession: Secondary succession.

Explanation: The area was not barren to begin with so primary succession is not correct. Secondary succession occurs after flooding and re-establishes a community where one previously existed.

Climax community: Paperbark scrub forest.

3

MAIN FEATURE IN FOREST SUCCESSION	NUMBER IN FIGURE
Grasses, perennials	3
Climax forest	6
Bare rock	1
Fast-growing trees	5
Woody pioneers	4
Mosses, grasses	2

4

	R-SELECTED SPECIES	K-SELECTED SPECIES
DEFINITION	Fast-growing and reproducing organisms such as pioneer plants.	Slow-growing, larger and long-lived organisms.
THE STAGE THEY APPEAR IN ECOLOGICAL SUCCESSION	They are often the first to occupy unused resources and living space.	Climax community.
WHY THEY ARE NORMALLY FOUND IN THE STAGE YOU HAVE IDENTIFIED	Pioneer species are normally small and photosynthetic. Effective seed dispersal, rapid growth, and rapid reproduction are characteristics they have that make them successful.	The community has become relatively stable and these species out-compete the others around them.
NAME OF THE POPULATION GROWTH CURVE	J-shaped, exponential growth curve.	S-shaped, logistic growth curve.

5 Primary succession occurs following the development of bare sites with no organisms inhabiting the affected area, for example catastrophic events such as volcanic eruptions, cyclones, earthquakes and tsunamis. In contrast, secondary succession is a response to recently disturbed community, for example, fire, tsunami, flood, or through human intervention by logging and land clearing for agriculture.

5.2 FOSSIL RECORDS

1 The ancient Aboriginal rock paintings located in the West Kimberley region depict the local plants and animals and how they lived. The changes in the content of the paintings over time give evidence of changing climate patterns. Fossils found in the area can provide further evidence of the type of climate and environment at the different times by inferring which types of plants and animals would grow in certain environments. Ice cores from the Antarctic could give evidence of a global ice age around 10 000 years ago.

2 Student answers can vary. An example of a response is below.

Merychippus and Pliohippus 23–2.6 million years ago. Shrubs and grassy plains. From 23 to 5 million years ago, there was an increase in the height of teeth crowns and horses became better at running, with an increase in both body size and length. Bones in the legs were fused so that the horses could stand on the tips of their toes (an adaptation for running). This indicates, with an even drier climate, trees were disappearing giving way to shrubs and grassy plains. Open prairie with tough grasses in a drier climate was the probable ecosystem type 5 to 2 million years ago evidenced by large teeth and body size as well as the loss of the side toes and one central toe (hoof) adapted to fast running.

3 It is because of the more than 250 sites at Riversleigh that are rich in fossils, many of them are well preserved. These sites are supplying a detailed and continuous fossil record of the changes in biota. The extensive biodiversity seen in the fossil record points to a climate very different to today's dry and hot habitat. Sedimentary rocks at the sites also indicate signs of a wetter climate. Scientists are using the characteristics of the grey limestone deposited between 25–15 million years ago. Early relatives of today's fauna were preserved in the lime-rich sediments of the wetlands that flourished at this time. This layer lies on top of older limestone without the fossil remnants and without sedimentation patterns characteristic of a wet climate.

5.3 HUMAN IMPACT ON BIODIVERSITY

1 Student answers will vary but can include the removal of tropical rainforests for agricultural purposes as an example of over-exploitation of resources, the introduction of the cane toad and its effect on populations of small mammals and native toads and frogs as an example of the introduction of new species and the fragmentation of populations due to habitat loss as an example of the disruption of ecological relationships.

2 With significantly reduced tree cover and increase in shallow-rooted grasses, the topsoil becomes more exposed to the abiotic elements, in particular to wind and rain. Compaction of the soil reduces rain infiltration and increased water pooling on the surface leading to water erosion. Tilling reduces topsoil causing nutrients and organic matter to be lost.

3

HUMAN IMPACT	MAGNITUDE	DURATION	SPEED
URBANISATION	Large	Very long, often permanent	Rapid
HABITAT DESTRUCTION	Large	Medium, depending on possibility of recolonisation in affected area	Rapid
LAND AND SOIL DEGRADATION	Large	If there is significant change or loss of topsoil, the ecosystem may be permanently changed.	Rapid
SALINITY	Large	Very long, often permanent	Medium
MONOCULTURE PRACTICES	Large	Permanent if crop is maintained	Rapid

4 Answers can include agricultural practices such as irrigation, use of fertilisers, pesticides and herbicides; introduced species; mining; burning fossil fuels.

CHAPTER 5 EVALUATION

1 a Primary succession is the colonisation of plants in a barren place. It describes the first biological occupation of a place where there were no living organisms previously. For example, the colonisation and the following succession of communities on a bare rock.

b Secondary succession is the recolonisation of disturbed plant communities.

2 a Examples include catastrophic events such as volcanic eruptions, cyclones, earthquakes and tsunamis.

b Examples include fire, flood, land clearing and other agricultural practices.

3 Slow-growing and long-lived species are found in climax communities. They are often forests with tall, well established trees.

4 Any four of: tolerant to extreme conditions, able to grow in poor quality soils with the ability to fix nitrogen from the atmosphere because of low nutrient levels, small and photosynthetic, effective seed dispersal, rapid growth, and rapid reproduction.

5 Trapped gas bubbles and the presence or absence of traces of organisms in sampled ice cores reveal information about the temperature and relative concentrations of atmospheric gases in past ecosystems. The past environment can be predicted using this information.

6 Some examples include: Aboriginal rock art in the area, properties of the rock the fossil was found in, ice cores from Antarctica dated to the same time as the fossil formation, fossil size and shape.

7 Some examples include agricultural practices such as monoculture, land clearing, use of pesticides, herbicides, fertilisers, mining, urbanisation, habitat fragmentation, irrigation, introduction of new species.

8 Possible impacts include habitat loss through natural vegetation removal and habitat fragmentation, which reduces living areas suitable for natural fauna and flora often leading to their extinction; loss of topsoil and nutrients as well as compacting of soil leading to erosion; waterlogging and salination.

9780170411745

CHAPTER 6 REVISION

6.1 CHROMOSOME STRUCTURE

1 In the nucleus of a eukaryotic non-dividing cell, DNA is only visible as a grainy substance without detail. Early microscopists named this seemingly diffuse, grainy substance chromatin. Chromatin is the cell's DNA together with all the proteins associated with it. When the cell prepares to divide, the chromatin condenses by coiling up. It eventually becomes thick enough to be seen, when stained, as a number of separate structures called chromosomes. A chromosome is one DNA molecule with its associated proteins.

For most of the time, chromosomes are unduplicated. An unduplicated chromosome is a single, long DNA double helix molecule coiled around histone proteins. By contrast, duplicated chromosomes, which have undergone DNA replication in preparation for cell division, contain two identical copies, called sister chromatids. Sister chromatids are joined by a centromere, making them appear as an 'X' shape.

6.2 PACKAGING OF DNA

1 In eukaryotes, DNA is found in a number of linear chromosomes in the nucleus and also in mitochondria and chloroplasts. The DNA of both mitochondria and chloroplasts is similar to the single, circular chromosomes in prokaryotic cells.

2 In the nucleus, the DNA double helix is packaged by special proteins, called histones, to form a complex called chromatin which looks grainy and diffuse. The chromatin undergoes further condensation to form the chromosome which is normally only visible during cell division.

3 Since chromosomes only become visible just prior to cell division and after replication has occurred, most images and drawings of chromosomes show them in the duplicated form.

4 A chromosome is a structure composed of DNA and protein that contains along its length linear arrays of genes carrying genetic information. By contrast, a gene is a unit of heredity that transmits information from one generation to the next.

5 **3** The loops of coiled DNA are wound around a core of eight histone proteins to produce a nucleosome.

2 Interacting proteins package loops of coiled DNA into a 'supercoil', to produce chromatin, which is organised as a cylindrical fibre.

5 The double helix of DNA is held together by hydrogen bonds between nitrogenous bases.

1 A tightly coiled and condensed human chromosome is only visible when stained during cell division, after DNA replication.

4 A nucleosome consists of a section of DNA molecule looped twice around a core of histones.

6.3 CASE STUDY: WHAT IS A GENE?

1 Pathogenic: disease causing.

2 Structure: a segment of DNA that codes for polypeptide. Function: a unit of heredity that transmits information from one generation to the next.

3 As a genome is all of the genetic material contained in an organism or a cell, it is much bigger than a gene and, in eukaryotes, it also includes the DNA in mitochondria and chloroplasts as well as the chromosomes within the nucleus.

4 The specific base pairing proposed by Watson and Crick was that C pairs with G and T with A. This suggested a copying mechanism for the genetic material because when the double stranded DNA separates and free nucleotides pair with the complementary exposed bases, a new strand of DNA will be formed when the nucleotides are joined together by an enzyme.

5 Griffith's observations with *S. pneumoniae* was the first time it had been shown that the hereditary structure that could be passed from one organism to another was heat stable. Using the same organisms, Avery's experiment showed that because this substance was broken down by a nuclease, it was a nucleic acid.

6.4 DNA AND ITS REPLICATION

1 **a** Each parent supplied half the chromosomes in a person. Either both strands of the DNA helix are maternal or both strands are paternal.

b The DNA of all organisms has the same form and uses the same genetic code, only differences in the base sequences account for the differences between organisms.

c Each chromosome is made of one DNA molecule.

d The different cell types found in a given individual's body contain identical DNA.

e In sexually reproducing organisms, all of the organism's body cells contain exactly the same DNA, half of which came from the mother and half from the father.

2 A molecule of DNA is composed of two long strands of subunits called nucleotides, wound around each other to form a shape like a twisted ladder or a spiral staircase, which is a double helix.

3 There are four kinds of nitrogenous (nitrogen containing) bases in DNA: adenine (A), thymine (T), guanine (G) and cytosine (C).

4 A nucleotide has three chemical components: a five-carbon sugar (deoxyribose in DNA), a phosphate group and a nitrogen-containing base.

5 The two strands are held together by hydrogen bonding between complementary nitrogenous bases. A bonds with T and C bonds with G (referred as the base-pairing rule).

6 Labels from the top: Left side: Deoxyribose-phosphate backbone, parental strand. Right side: parental helix, replicate strand.

7 The double helix is separating into its two individual strands. Two new strands are being made alongside these, resulting in two double strands of DNA, identical to the original molecule being produced. This is called DNA replication.

8 Within the nucleus, stockpiles of free nucleotides attach to the exposed bases according to the base-pairing rule.

9 DNA replication is considered to be 'semi-conservative' because the outcome is two double helix molecules, each consisting of one parental strand and one new strand.

10 The junction between the unwound single strands of DNA and the intact double helix is the replication fork.

11 The enzyme DNA helicase unzips a small section of the double-stranded DNA by breaking the weak hydrogen bonds between the nucleotides and exposing the nucleotide bases. The replication fork, between the unwound single strands of DNA and the intact double helix, moves along the DNA so that there is a continuous unwinding of the parental strands. Free nucleotides attach to the exposed bases according to the base-pairing rule and the enzyme DNA polymerase joins the nucleotides to form a new complementary strand.

CHAPTER 6 EVALUATION

1 B

2 C

3 D

4 During replication, one original parental strand forms one new strand. This is called semi-conservative because one of the two strands is conserved, or retained, from one generation to the next, while the other strand is new.

5 The enzyme DNA helicase unzips a small section of the double-stranded DNA by breaking the weak hydrogen bonds between the nucleotides and exposing the nucleotide bases. By contrast, the enzyme DNA polymerase, joins the nucleotides that have bonded to the exposed bases to form a new complementary strand.

6 DNA structure to enable transmission: double helix with sugar-phosphate backbone, subunits are nucleotides, complementary nitrogenous bases form rungs, hydrogen bonds holds bases together, A with T and G with C. DNA replication: DNA helicase unwinds and separates strands, free nucleotides pair with complementary exposed bases, DNA polymerase joins nucleotides to form new strands, called semi-conservative replication as in the two identical DNA molecules formed, one strand is original and one strand is new.

CHAPTER 7 REVISION

7.1 KARYOTYPES, MEIOSIS AND HYBRIDISATION

1 Microscopic examination of a stained eukaryotic cell in the process of nuclear division reveals a jumbled cluster of chromosomes that exist in pairs, called homologous pairs. The exception is the sex chromosomes which in one sex, usually the male, are not homologous. One chromosome of each homologous pair is inherited from each parent, with members of each homologous pair sharing characteristic banding patterns. A karyotype is the standard presentation form used to display and analyse chromosomes. Photographic images of chromosomes, during cell division, are arranged into matched and ordered pairs to create a karyotype. The chromosomes are ordered by length, from largest to smallest.

2 Gametes contain half the number of chromosomes found in body cells.

3 Meiosis is appropriate for gamete formation because it halves the number of chromosomes in the resulting daughter cells. This means the chromosome number can be restored by fertilisation.

4 The karyotypes of sperm cells would have half the number of chromosomes as the male they came from, and half the sperm karyotypes would contain one X chromosome and the other half would contain one Y chromosome.

5 Complete the following table of chromosome numbers in various species.

SPECIES	2n	NUMBER OF HOMOLOGOUS CHROMOSOME PAIRS	n
Human	46	23	23
Fruit fly	8	4	4
House fly	12	6	6
Chimpanzee	48	24	24
Camel	70	35	35
Chicken	78	39	39
Goat	60	30	30
Petunia	14	7	7
Rice	24	12	12

6 a The cells in the donkey with 31 chromosomes would be the sex cells – eggs and sperm.

b Somatic cells (non-sex cells) in the zebra would contain 44 chromosomes.

c There would be 62 chromosomes in the somatic cells of the donkey.

d The 2n number of a 'zonkey' would be 53.

e A karyotype is the standard presentation form used to display and analyse chromosomes. Photographic images of chromosomes, during cell division, are arranged into matched and ordered pairs to create a karyotype. The chromosomes are ordered by length, from largest to smallest.

f A zebra karyotype would contain 22 pairs of chromosomes, whereas a zonkey karyotype would contain 22 unpaired zebra chromosomes and 31 unpaired donkey chromosomes.

g When the zonkey produces gametes, the 53 chromosomes will not be distributed evenly into the gametes. Some gametes are likely to have more chromosomes than others.

h Most hybrid animals are infertile because the chromosomes are not homologous pairs and are unable to match up in metaphase of meiosis I, therefore cannot be evenly distributed into the gametes.

7.2 APHIDS: AN UNUSUAL REPRODUCTIVE STRATEGY

1 The eggs produced in autumn are diploid.

2 Both eggs and sperm are produced by meiosis, and both would carry one X chromosome.

3 Overwintering eggs only hatch into females because they are all XX (female).

4 In humans and aphids, all eggs carry an X chromosome. Half the sperm in humans carry an X chromosome and half carry a Y chromosome, so that half the offspring will be female and half male. In aphids, all sperm carry an X chromosome, so all offspring will be female (XX)

5 If sperm without a sex chromosome fertilised eggs, the eggs would develop into males. It is more beneficial for aphids to only have females in spring, because food is readily available and asexual parthenogenic reproduction can increase the population size much faster than sexual reproduction.

7.3 OOGENESIS AND SPERMATOGENESIS

1 The process of oogenesis begins in the ovaries of females during embryonic development, before a woman is born. There, primary oocytes begin meiosis, but remain in prophase I until the female matures sexually. After that time, a primary oocyte completes meiosis I each month to form a secondary oocyte and a structure called a polar body. Cytokinesis is unequal with almost all of cytoplasm going into the secondary oocyte. The polar body degenerates. The second meiotic division, which produces a haploid ovum (egg) and a second polar body, occurs only if a sperm fertilises the egg.

The production of sperm in males is called spermatogenesis. Stem cells in the testes undergo mitotic division, each time producing a new stem cell that continues to divide, and a diploid primary spermatocyte. The latter divides in meiosis I to form two secondary spermatocytes which in turn divide in meiosis II to form four spermatids, which are haploid and develop into four sperm cells. This process occurs throughout a male's lifetime and is capable of producing, in humans, at least 3 million sperm per day.

In the process of fertilisation, haploid sex cells fuse to produce a diploid zygote. Human gametes produced by meiosis each contain n = 23 chromosomes. Fertilisation restores the chromosome number to 2n = 46. Different species have different numbers of chromosomes.

7.4 CASE STUDY: A UNIQUE SYSTEM OF SEX DETERMINATION

1 In humans, all eggs contain one X chromosome and half the sperm contain an X and half contain a Y chromosome. If the zygote is XX it is a female and if XY it is male.

2 During synapsis in prophase I, when the pairs of homologous chromosomes coil around each other, non-sister chromatids attached at points called chiasmata, where they exchange segments of genetic material with one another in a process called crossing over. This recombination shuffles pieces of maternal and paternal genes and rearranges the combinations of alleles available on each homologous chromosome.

3 The recombinant chromosomes that result from crossing over contain new combinations of genes which increases the genetic diversity in the offspring.

4 In metaphase I, when chromosomes line up in homologous pairs across the equator of the cell, each pair of maternal and paternal chromosomes lines up independently of other pairs. The result is that the original maternal and paternal chromosomes are distributed randomly to the gametes instead of as a pre-defined set from either parent.

5 Independent assortment during meiosis shuffles existing alleles into different combinations which greatly increases the variation in the genotypes of the offspring.

6 Sex determination by temperature would encourage out breeding and prevent inbreeding because all the offspring in a clutch would be the same gender and so would not be able to mate with each other. They would need to find a mate from another clutch, which would increase variation in the population.

7 The karyotype of a human male has two different sex chromosomes – an X and a Y. The karyotype from a crocodile would not have different sex chromosomes, as their sex depends on their environment, not on their chromosomes.

CHAPTER 7 EVALUATION

1 B

2 D

3 B

4 Asexual reproduction is carried out by one parent; therefore, the offspring closely resemble that parent. As sexual reproduction requires two sources of hereditary material, carried in gametes and usually from different parents, it generates much more genetic variation in the offspring.

5 In oogenesis, cytokinesis in both meiosis I and II is unequal and produces only a single egg and small polar bodies that degenerate. By contrast, meiosis in spermatogenesis produces four sperm. At birth an ovary contains all the cells that will ever develop into eggs. Once puberty is reached, sperm are produced throughout a man's lifetime. Sperm are produced continuously, whereas oogenesis has long breaks between stages of meiosis – up to 40 years.

6 **a** In humans, crossing over occurs in the nucleus during synapsis in prophase I of meiosis.

b Crossing over increases variation in offspring.

c When pairs of homologous chromosomes coil around each other to form a bivalent, non-sister chromatids become attached at chiasmata. Here they exchange segments of genetic material, generating new combinations of alleles on homologous chromosomes. This increases the genetic diversity in the offspring.

CHAPTER 8 REVISION

8.1 GENOME AND GENES

1 Nucleotide bases, genes, DNA strands, genome

8.2 PROTEIN SYNTHESIS

1 1U, 2G, 3C, 4A, 5U, 6C, 7C, 8G, 9C

2 13A, 14C, 15G, 16U, 17A, 18G, 19G, 20C, 21G

3 10U, 11A, 12A or 10U, 11A, 12G or 10U, 11G, 12A

4 22 arginine, 23 isoleucine, 24 cysteine

5 A transcription, nucleus

B translation, at ribosome in cytoplasm

8.3 CODING AND NON-CODING DNA

1 **a** The small segments of DNA, found within mRNA exons, that are used as templates for mRNA synthesis and thus for polypeptide synthesis; also known as genes.

b All DNA sequences within the genome that are not found within mRNA-coding exons, that is, do not code for polypeptides.

2 Functions include: transcribed into functional RNA, including ribosomal RNA (rRNA) and transfer RNA (tRNA), both of which are necessary for protein synthesis; regulatory DNA, involved in producing proteins for switching genes on and off; centromeres – hold chromatids of chromosomes together in cell division; telomeres – DNA sequences that protect coding DNA sequences by being 'sacrificed/lost' off ends of DNA strands during DNA replication.

9780170411745

8.4 THE PURPOSE OF GENE EXPRESSION

1 Specialised cell types become so different to each other, and different within types at different times, because they don't express all the genes of their genome at any particular point in time. Rather, cells must have some genes that code for specific required proteins active, and other genes for unnecessary proteins inactive.

2 a A gene being transcribed into mRNA and translated into a polypeptide

b Various processes that enable a gene to be expressed (or not) in specific cells at specific times

3 a F

b T

8.5 THE FACTORS REGULATING THE PHENOTYPIC EXPRESSION OF GENES

1

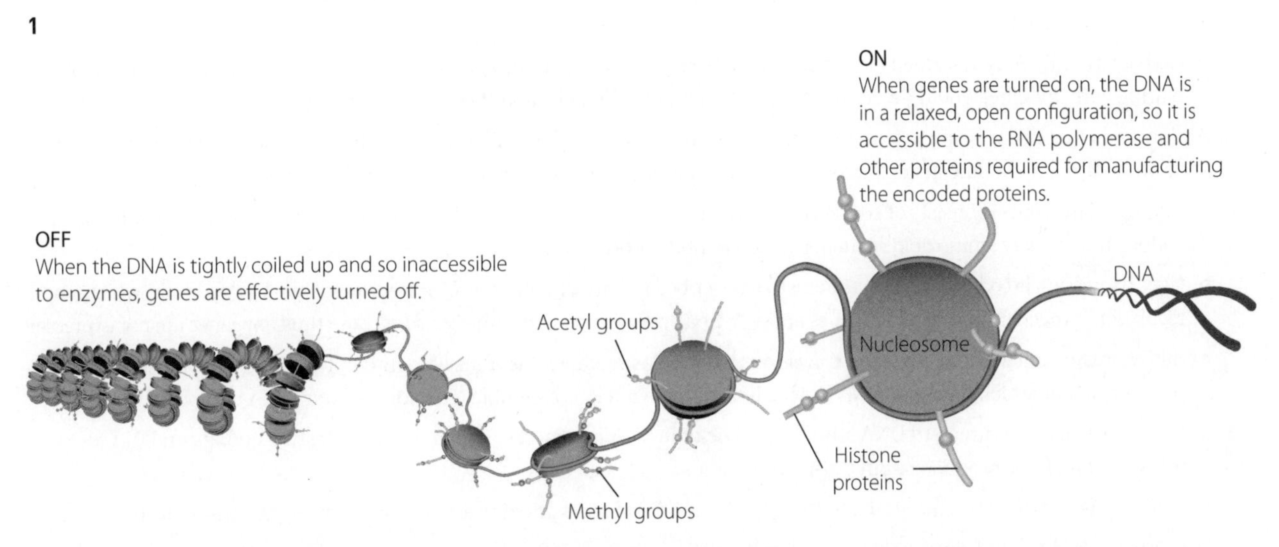

When DNA is packaged and coiled into chromatin, the genes within the DNA are not available for expression. The DNA strand is wrapped and coiled tightly into a very condensed structure, containing nucleosomes, in which the DNA strand encircles histone proteins, and then the nucleosomes are further spiralled. RNA polymerase is unable to access the DNA nucleotides in the chromatin to begin transcription of the mRNA molecule from the DNA template strand. In essence, such genes are 'switched off' and the corresponding polypeptides are not made.

2-3 Student answers will vary, but should contain terms as required in question, but may contain different connections.

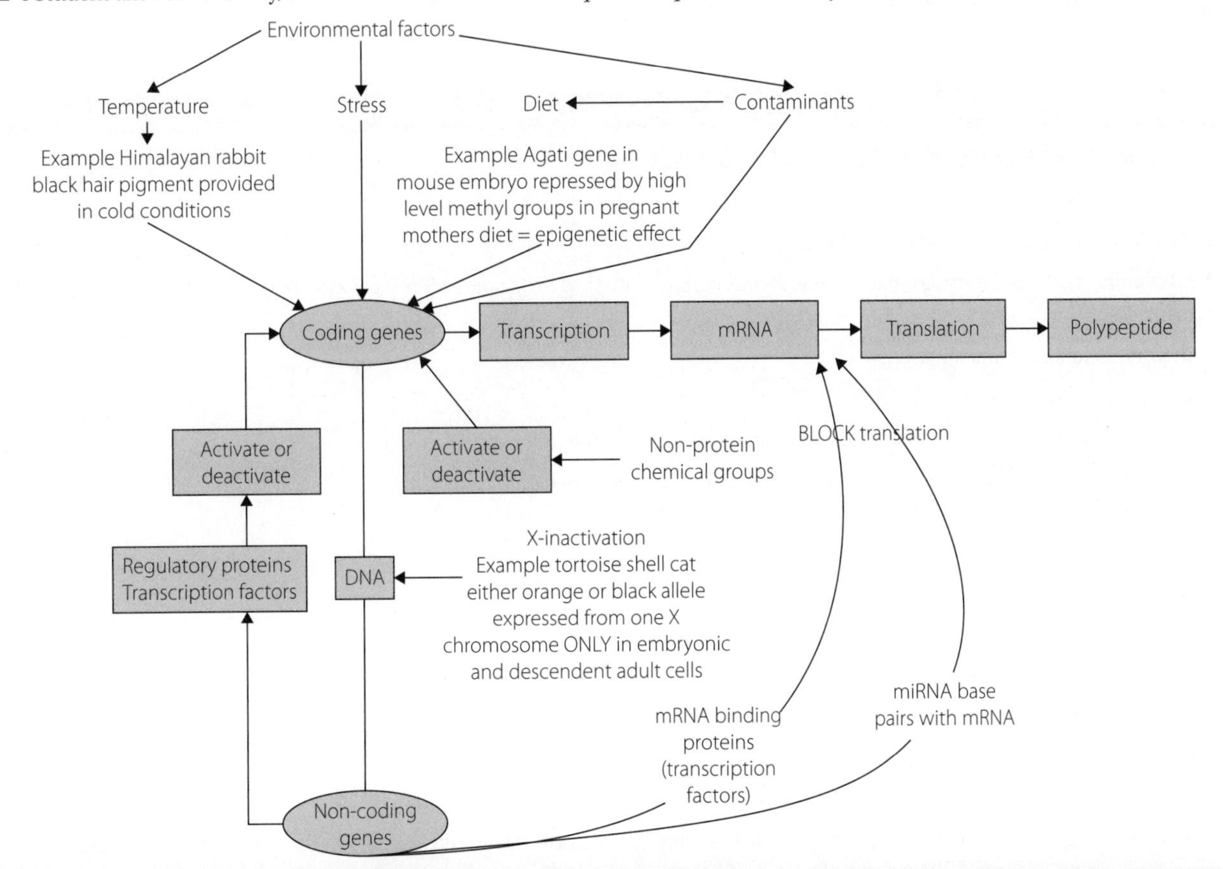

8.6 TRANSCRIPTION FACTORS REGULATE MORPHOLOGY AND CELL DIFFERENTIATION.

1 Terms in order: morphology, embryonic, differentiation, fertilisation, proteins, zygote, expression, structures, sex, male, regulatory, activate.

CHAPTER 8 EVALUATION

1 A

2 B

3 a The order of information transfer is: DNA codons to mRNA codons (used in genetic code) to tRNA anticodons to amino acids. Go back to mRNA codon (that attracted tRNA containing anticodon) and use genetic code to look up amino acid corresponding to that mRNA codon.

b Go back to mRNA as described in part a, and the base pair sequence complementary to this mRNA (and with thymine pairing with adenine) would determine the DNA base pair sequence and therefore codon sequence

4 Amino acids that occur frequently in proteins that are required in high numbers, have multiple codons capable of directing their production, but amino acids that occur irregularly in proteins would have one codon for their production.

5 a Coding DNA (genes) consists of codons from which polypeptides will be produced, noncoding DNA does not consist of codons that specify amino acid sequence in polypeptide production

b Gene expression is codons in DNA are actually transcribed into mRNA and mRNA is translated into a polypeptide, while gene regulation is mechanisms by which genes are 'switched on' for expression, or 'switched off' and thus not genes are not expressed.

c mRNA is transcribed from DNA and travels out to cytoplasm where it is translated into polypeptide by a ribosome/ ribosomes that attach, while in rRNA the nucleolus in the nucleus assembles ribosomes from rRNA and proteins.

6 a Regulatory proteins attach to DNA strand and block the RNA polymerase enzyme from accessing codons on DNA for mRNA strand to be produced from DNA template

b Similarly, regulatory proteins attach to the mRNA strand and block the ribosomes from carrying out translation

7 Students could describe hair pigmentation change in Himalayan rabbit, or agouti gene repression in mice embryos due to mother's diet containing certain level of methyl chemical groups.

Cold conditions over certain period of time induce black pigmented hair instead of white being produced by cells exposed to the lowered temperatures: pigment production by gene expression is regulated by temperature. Methylation of agouti gene causes it to be 'switched off' so that it is not expressed to produce agouti phenotype characteristics in offspring of pregnant mothers having consumed food with certain high levels of methyl groups. The passing on of regulated genes from one generation to the next is an epigenetic effect.

CHAPTER 9 REVISION

9.1 IDENTIFYING ERRORS IN GENES AND CHROMOSOMES

1 G-T-T-T-A-C-T-G-G-C-C-A-A-G-T-A-G-G

2 C-A-A-A-U-G-A-C-C-G-G-U-U-C-A-U-C-C

3 Glutamine, methionine, threonine, glycine, serine, serine. Note that AUG acts as START codon only if specific initiating factors are present.

1 a point deletion, frame-shift

b C-A-A-U-G-A-C-C-G-G-U-U-C-A-U-C-C-

c glutamine STOP (no further amino acids would be added to the polypeptide chain)

d probably huge effect: 5 amino acids following glutamine, as shown in Activity 1 part 3 answer, would be missing along with any subsequent amino acids coded for remainder of sequence

2 a point, substitution

b C-A-A-A-U-G-A-C-C-G-G-U-U-C-A-U-C-G

c glutamine, methionine, threonine, glycine, serine, serine

d no effect as UCC and UCG on mRNA both code for serine

3 a point, substitution

b C-A-A-A-U-U-A-C-C-G-G-U-U-C-A-U-C-C

c glutamine, isoleucine, threonine, glycine, serine, serine

d One wrong amino acid may affect polypeptide function if the substituted amino acid has very different properties to original

4 **a** point insertion, frame-shift

b C-A-A-A-U-G-A-C-C-G-G-U-U-C-A-U-U-C-C-

c glutamine, methionine, threonine, glycine, serine, phenylalanine

d Probably huge effect: after serine (5th amino acid in sequence), phenylalanine and every amino acid following it will be different to those originally coded for

9.2 ANEUPLOIDY AND KARYOTYPE

1

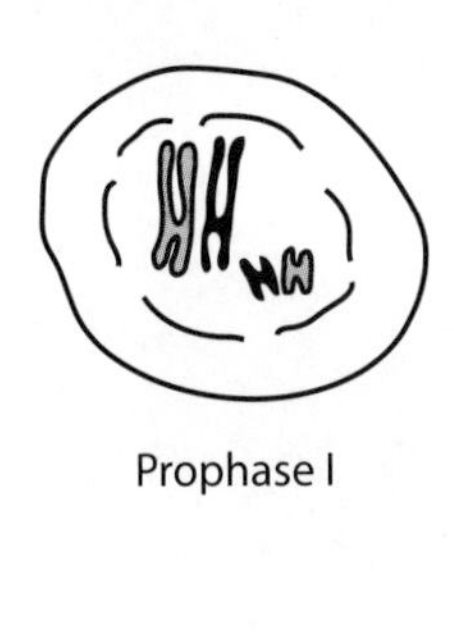
Prophase I

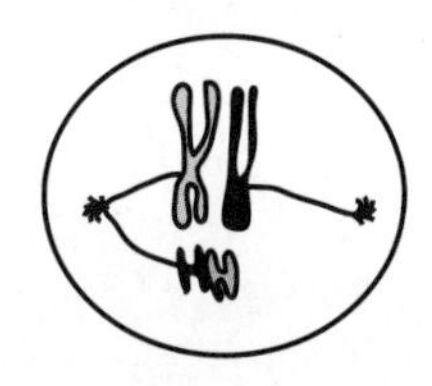
Metaphase I

Anaphase I

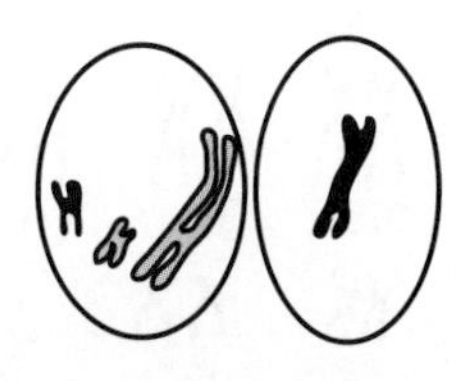
Telophase I

Prophase II

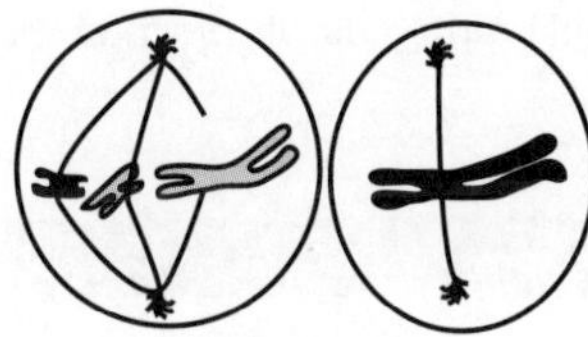
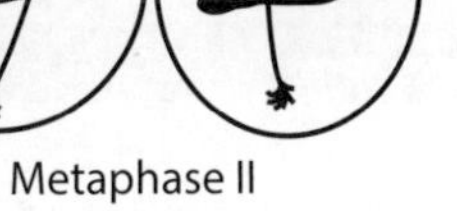
Metaphase II

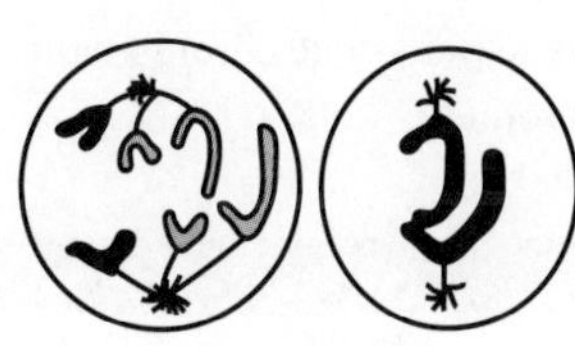
Anaphase II

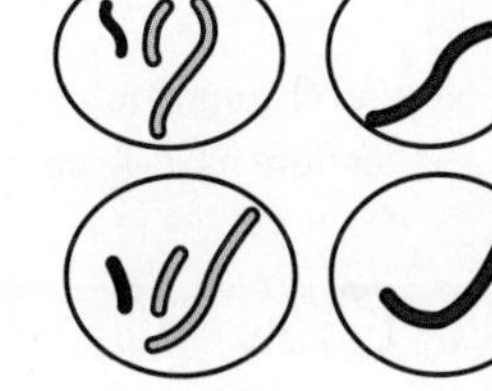
Telophase II

2

Prophase I

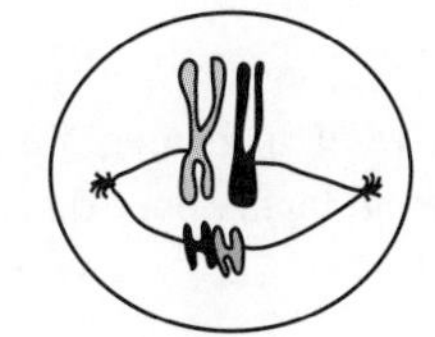
Metaphase I

Anaphase I

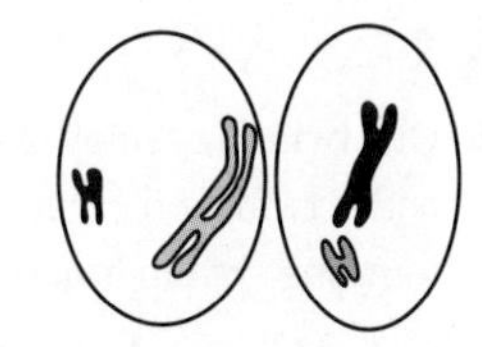
Telophase I

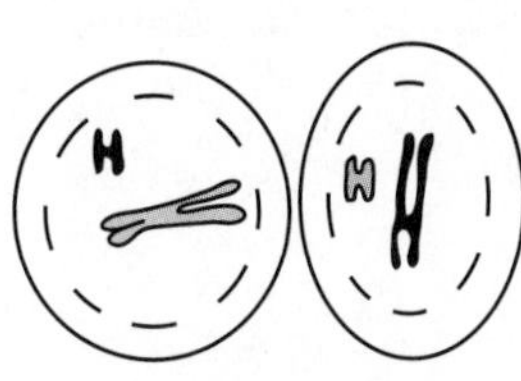

Prophase II

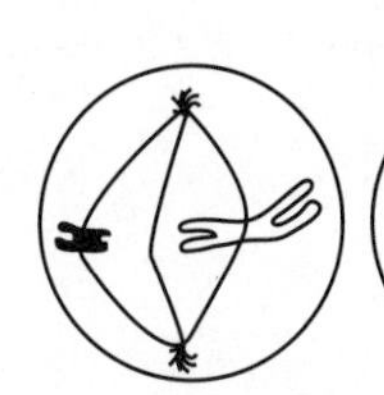
Metaphase II

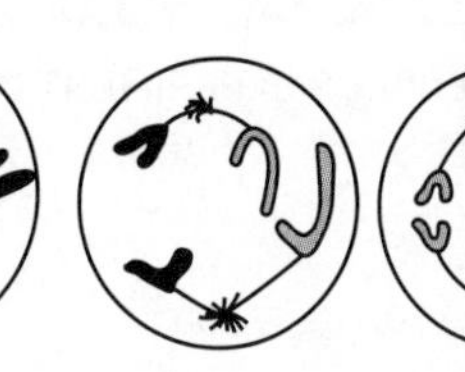
Anaphase II

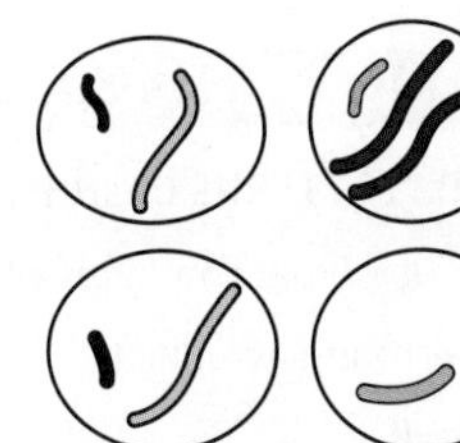
Telophase II

1

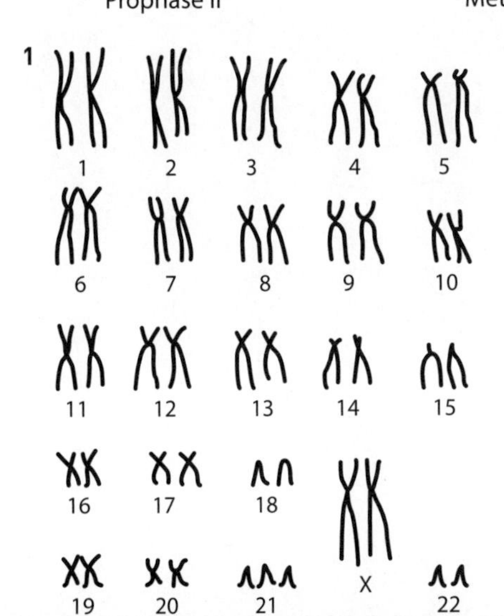

2 Chromosome 21 is quite small with a relatively small number of genes. Repetition of chromosomes with many more genes, such as numbers 1 or 2, cause too many problems for the embryos to survive.

1

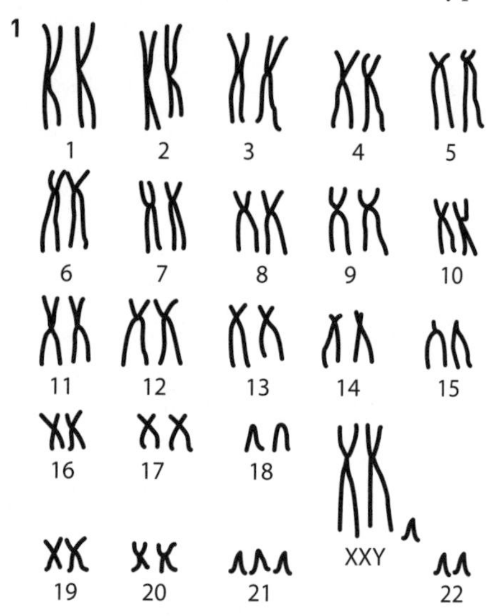

2 Normal chromosome set has 23 pairs or 46 chromosomes. People with Klinefelter's syndrome have an extra X chromosome, so therefore have 47 chromosomes.

CHAPTER 9 EVALUATION

1 C

2 B

3 Length, position of centromere, banding pattern

4 X + XY; XX + Y

5 One twin may enjoy and play more sport than the other. This would result in greater muscular development of the more active twin, and other possible health and lifestyle benefits, for example, higher bone density, better sleep patterns, better immune system function.

One twin may enjoy and eat different foods to the other – the more nutritious diet could result in improved growth (and development, dependent on age), healthy weight maintenance, better functioning immune system (better resistance to infection).

CHAPTER 10 REVISION

10.1 PATTERNS OF INHERITANCE: 'MENDEL'S EXPERIMENTS'

1 Autosomal dominant allele

2 **a** round, wrinkled

b R, r

c RR, rr

d Rr

e

F1 cross (to predict F2 genotypes)		F1 parent gametes	
		$\frac{1}{2}$ R	$\frac{1}{2}$ r
F1 parent Gametes	$\frac{1}{2}$ R	$\frac{1}{4}$ RR round	$\frac{1}{4}$ Rr round
	$\frac{1}{2}$ r	$\frac{1}{4}$ Rr round	$\frac{1}{4}$ rr wrinkled

f $\frac{3}{4}$ round : $\frac{1}{4}$ wrinkled

9780170411745

10.2 PATTERNS OF INHERITANCE: AUTOSOMAL DOMINANT ALLELES

1 (I) Punnet square:

P cross: BB × Bb		parent 1 gametes	
		$\frac{1}{2}$ B	$\frac{1}{2}$ B
parent 2 gametes	$\frac{1}{2}$ B	$\frac{1}{4}$ BB black	$\frac{1}{4}$ BB black
	$\frac{1}{2}$ b	$\frac{1}{4}$ Bb black	$\frac{1}{4}$ Bb black

phenotypic F1 frequencies: all black

(II) Punnet square:

P cross: Bb × bb		parent 1 gametes	
		$\frac{1}{2}$ B	$\frac{1}{2}$ b
parent 2 gametes	$\frac{1}{2}$ b	$\frac{1}{4}$ Bb black	$\frac{1}{4}$ bb brown
	$\frac{1}{2}$ b	$\frac{1}{4}$ Bb black	$\frac{1}{4}$ bb brown

phenotypic F1 frequencies: $\frac{1}{2}$ black : $\frac{1}{2}$ brown

(III) Punnet square:

P cross: Bb × Bb		parent 1 gametes	
		$\frac{1}{2}$ B	$\frac{1}{2}$ b
parent 2 gametes	$\frac{1}{2}$ B	$\frac{1}{4}$ BB black	$\frac{1}{4}$ Bb black
	$\frac{1}{2}$ b	$\frac{1}{4}$ Bb black	$\frac{1}{4}$ bb brown

phenotypic F1 frequencies: $\frac{3}{4}$ black : $\frac{1}{4}$ brown

2 Probable parent genotypes: Ss and ss

P cross: Ss × ss		parent 1 gametes	
		$\frac{1}{2}$ S	$\frac{1}{2}$ s
parent 2 gametes	$\frac{1}{2}$ s	$\frac{1}{4}$ Ss spiny	$\frac{1}{4}$ ss smooth
	$\frac{1}{2}$ s	$\frac{1}{4}$ Ss spiny	$\frac{1}{4}$ ss smooth

resulting F1 frequencies: $\frac{1}{2}$ spiny : $\frac{1}{2}$ smooth

10.3 PATTERNS OF INHERITANCE: INCOMPLETE AND CO-DOMINANCE

1 $D^0D^0 \times D^1D^1$; $D^0D^0 \times D^1D^0$; $D^1D^0 \times D^1D^0$

2 **a** $D^0D^0 \times D^1D^1$ would produce the most palomino offspring

P cross: $D^0D^0 \times D^1D^1$		D^0D^0 parent gametes
		all D^0
D^1D^1 parent gametes	all D^1	all D^1D^0 all palomino

b highest frequency of palomino offspring = 100% = all palomino

1 Symbols: black C^B, white C^W

2 Genotypes of the 3 phenotypes: black C^BC^B, white C^WC^W, erminette C^BC^W

P cross: $C^BC^B \times C^WC^W$		C^BC^B parent gametes
		all C^B
C^WC^W parent gametes	all C^W	all C^BC^W all erminette

3 (A) P: black × white

F_1 expected phenotypic frequency: all erminette

(B) P: black × erminette

P cross: $C^BC^B \times C^BC^W$		C^BC^B parent gametes
		all C^B
C^BC^W parent gametes	$\frac{1}{2}C^B$	$\frac{1}{2}C^BC^B$ black
	$\frac{1}{2}C^W$	$\frac{1}{2}C^BC^W$ erminette

F_1 expected phenotypic frequency: $\frac{1}{2}$ black: $\frac{1}{2}$ erminette

(C) P: erminette × erminette

P cross: $C^BC^W \times C^BC^W$		C^BC^W parent gametes	
		$\frac{1}{2}C^B$	$\frac{1}{2}C^W$
C^BC^W parent gametes	$\frac{1}{2}C^B$	$\frac{1}{4}C^BC^B$ black	$\frac{1}{4}C^BC^W$ erminette
	$\frac{1}{2}C^W$	$\frac{1}{4}C^BC^W$ erminette	$\frac{1}{4}C^WC^W$ white

F_1 expected phenotypic frequency: $\frac{1}{4}$ black: $\frac{1}{2}$ erminette: $\frac{1}{4}$ white

9780170411745

10.4 PATTERNS OF INHERITANCE: SEX-LINKED INHERITANCE

1 Females: $X^C X^C$ or $X^C X^c$ or $X^c X^c$; males: X^CY or X^cY

2 Colour blindness: $X^c X^c$ or X^cY

3 (I) P: $X^C X^C \times X^cY$ = homozygous normal female X colour blind male

P cross:		X^CX^C female parent gametes
		all X^C
X^cY male parent gametes	$\frac{1}{2}$ X^c	$\frac{1}{2}$ X^CX^c normal (carrier) female
	$\frac{1}{2}$ Y	$\frac{1}{2}$ X^CY normal male

F_1 expected phenotypic frequency: $\frac{1}{2}$ normal (carrier) female: $\frac{1}{2}$ normal male

(II) P: $X^c X^c \times X^CY$ = colour blind female X normal male

P cross:		$X^c X^c$ female parent gametes
		all X^c
X^CY male parent gametes	$\frac{1}{2}$ X^C	$\frac{1}{2}$ X^CX^c normal (carrier) female
	$\frac{1}{2}$ Y	$\frac{1}{2}$ X^cY colour blind male

F_1 expected phenotypic frequency: $\frac{1}{2}$ normal (carrier) female: $\frac{1}{2}$ colour blind male

(III) $X^C X^c \times X^cY$ = normal (carrier) female X colour blind male

P cross:		$X^C X^c$ parent gametes	
		$\frac{1}{2}$ X^C	$\frac{1}{2}$ X^c
X^cY parent gametes	$\frac{1}{2}$ X^c	$\frac{1}{4}$ X^CX^c normal (carrier) female	$\frac{1}{4}$ X^cX^c colour blind female
	$\frac{1}{2}$ Y	$\frac{1}{4}$ X^CY normal male	$\frac{1}{4}$ X^cY colour blind male

F_1 expected phenotypic frequency: $\frac{1}{4}$ normal (carrier) female: $\frac{1}{4}$ colour blind female: $\frac{1}{4}$ normal male: $\frac{1}{4}$ colour blind male

(IV) $X^C X^c \times X^CY$ = normal (carrier) female X normal male

P cross:		$X^C X^c$ parent gametes	
		$\frac{1}{2}$ X^C	$\frac{1}{2}$ X^c
X^cY parent gametes	$\frac{1}{2}$ X^C	$\frac{1}{4}$ X^CX^C normal female	$\frac{1}{4}$ X^CX^c normal carrier female
	$\frac{1}{2}$ Y	$\frac{1}{4}$ X^CY normal male	$\frac{1}{4}$ X^cY colour blind male

F_1 expected phenotypic frequency: $\frac{1}{4}$ normal female: $\frac{1}{4}$ normal carrier female: $\frac{1}{4}$ normal male: $\frac{1}{4}$ colour blind male

10.5 PATTERNS OF INHERITANCE: MULTIPLE ALLELES

1 **a** each parent's genotype: I^AI^A or I^Ai

b baby genotype: ii

c parents must both have had I^Ai genotype and both passed on the recessive i allele

P cross:		I^Ai parent 1 gametes	
		$\frac{1}{2}$ I^A	$\frac{1}{2}$ i
I^Ai parent 2 gametes	$\frac{1}{2}$ IA	$\frac{1}{4}$ I^AI^A A blood group	$\frac{1}{4}$ I^Ai A blood group
	$\frac{1}{2}$ i	$\frac{1}{4}$ I^Ai A blood group	$\frac{1}{4}$ ii O blood group

Although both of you parents have A type blood, you are carrying a factor, called a recessive factor 'i', that is 'masked' by your A blood factor. Both of you passed this recessive i factor on to your baby, and the double occurrence of this factor produces O group blood. There is a 1 in 4 chance that you will produce a baby with O group blood each time you conceive, and a 3 in 4 chance that your baby will have A group blood.

10.6 POLYGENIC INHERITANCE

1 E, C, D, B, A

CHAPTER 10 EVALUATION

1 B

2 B

3 **a** (i) both black; (ii) black and brown; (iii) both black

b both parents must be homozygous (pure breeding) brown: bb

c

P cross: bb X bb		parent 1 gametes all b
parent 2 gametes	all b	all bb brown

4 correct insertions: differences; both sons and daughters

9780170411745

CHAPTER 11 REVISION

11.1 FUNDAMENTALS OF BIOTECHNOLOGY

1

STRUCTURE OR SUBSTANCE	DESCRIPTION OF ACTION
Plasmid	Small circular self-replicating DNA molecule
Gel electrophoresis	Sorts DNA molecules based on size and charge
DNA ligase	An enzyme that joins two segments of DNA together
Restriction site	Specific site at which restriction enzymes cut DNA
Taq polymerase	An enzyme used in PCR that catalyses the synthesis of DNA
Vector	Vehicle to introduce DNA into a host cell
DNA polymerase	An enzyme that catalyses the synthesis of DNA
Blunt ends	Results from cleavage by a restriction enzyme in the middle of the recognition sequence
Primer	Synthetic short, single-stranded DNA molecule

2 Plasmids are extracted from bacteria by rupturing the cell membranes and cell walls. Similarly, the DNA of interest is isolated from the donor organism. The same restriction enzyme is used to cut the plasmid DNA and the DNA of the gene to be inserted, to ensure they have complementary sticky ends. The plasmid vectors and the gene of interest are mixed together and their sticky ends pair. DNA ligase is used to join the two segments to form recombinant plasmids. These plasmids are added to a bacterial culture, where they are taken up by some bacteria in a process called transformation. When the bacteria reproduce by dividing, the plasmid is also replicated. This generates numerous copies of the recombinant DNA. A process called antibiotic selection can be used to identify transformed bacteria.

3

NAME OF PROCESS	CONDITIONS FOR PROCESS	ORDER	PURPOSE
Extension	temperature raised to 72°C	3	new DNA strands are synthesised starting from primers
Denaturation	temperature raised to 95°C	1	hydrogen bonds between bases in double-stranded DNA broken to separate strands
Annealing	temperature reduced to 50–60°C	2	primers join to complementary sequences on DNA to be copied

11.2 USING GEL ELECTROPHORESIS FOR GENETIC TESTING

1 Restriction enzymes are enzymes that cuts DNA at a specific restriction site.

2 Cleavage of DNA by a restriction enzyme may form overhanging steps, called sticky ends which leave some nucleotides exposed.

3 Electrophoresis makes use of the negative charge of DNA to separate DNA fragments by size within an agarose gel. When an electric current runs through the gel, DNA fragments move towards the positive electrode. The gel acts as a sponge, with smaller fragments that experience less resistance migrating through the cross-linked gel matrix faster and further than the larger fragments.

4 DNA itself is not visible in the gel. To view the separated DNA fragments, the gel is stained with ethidium bromide. The dye binds to DNA and fluoresces under ultraviolet light, showing a pattern of bands that can then be photographed. Each band on the gel contains millions of pieces of DNA of the same size.

5 **a** The person with sickle cell anaemia is labelled b.

b An individual without sickle cell anaemia is labelled c.

c The carrier of sickle cell anaemia is labelled a.

11.3 DNA PROFILING IN THE STUDY OF REPRODUCTIVE STRATEGIES

1-4

FAMILY	BIOLOGICAL CYGNETS	PARASITIC CYGNETS
1	C1 C2	0
2	C1 C2 C4	C3 C5 2/5
3	C1 C3 C6	C2 C4 C5 3/6
4	C1 C2	0
5	C1 C2 C3	0
6	C1 C2 C3	0
7	C1 C4 C5	C2 C3 C6 C7 4/7
8	C1 C2	0

5 Brood parasitism occurred in families 2, 3, and 7.

6 The maximum proportion of parasitic cygnets in this sample was 9 out of 30.

7 The results confirm the belief that large Black Swan families are due to brood parasitism, because in all of the 8 families there were only 2 or 3 cygnets in each nest that could have been the biological offspring of the mother.

8 To determine whether a cygnet has been fathered by a male other than its social father, consider the banding pattern of each of the cygnets known to be related to their mother. One band should be able to be traced to the mother and one to the father. If this is not the case, then the social father is not the biological father. In Family 5, some bands of cygnet 5 do not match those of the social father or the mother.

CHAPTER 11 EVALUATION

1 C

2 C

3 B

4 A

5 A baby shares 50% of the bands in their DNA profile with their mother. This is because the baby inherits half their DNA from their mother and half from their father.

6 a Unwanted DNA in the PCR will be amplified, along with the target DNA. Contamination of this type could mean that the results of the study are misleading. For example, a DNA profile may give a false positive match or contaminant DNA may be mistaken for ancient DNA.

b Contamination could come from the person conducting the PCR or from the environment of the laboratory or from where the DNA was collected.

c To prevent contamination from occurring, gloves, masks and lab coats should be worn by people handling the specimens and special care must be taken during collection of samples.

CHAPTER 12 REVISION

12.1 DESCRIBING EVOLUTION

	CONCEPT	DEFINITION
1	Phylogenetic relationships	evolutionary relationships that exist between individuals or groups of organisms
2	Comparative genomics	the process of contrasting DNA sequences in different organisms
3	Molecular homology	when the common ancestry of organisms is identified from shared biomolecular elements used to test their relationships
4	Microevolution	small scale variation of allele frequencies within a species or population, in which the descendant is of the same taxonomic group as the ancestor
5	Molecular phylogeny	the study the evolutionary relationships between organisms by using DNA data
6	Evolutionary radiation	an increase in taxonomic diversity or morphological disparity over time
7	Macroevolution	variation of allele frequencies at or above the level of species resulting in the descendant being in a different taxonomic group to the ancestor

9780170411745

12.2 DNA VARIATION

1 The DNA of mitochondria and chloroplasts are similar to the single circular chromosomes of prokaryotes. They code for many of the proteins specific for the function of these organelles.

2 Next generation sequencing methods are much more powerful, less expensive and faster than those used in the Human Genome Project.

3 The genomes of organisms can be used to determine their evolutionary relationships, because as the degree of similarity of the nucleotide sequences in two organisms increases, so does the closeness of the relationships between those species.

4 The difference in DNA sequences between the genomes of the cow and sheep are 7.5% compared with 20.0% between that of the cow and the pig. This shows that sheep are more closely related to cows than pigs.

5 The most distantly related primate to humans is the African galago with a difference of 28.0%. By contrast, the most closely related primate to humans would be the chimpanzees, with a difference of only 1.6%; they would have diverged most recently. Humans and gibbons are the next most closely related (3.5%) and then humans and rhesus monkey (5.5%). This information shows that humans and chimpanzees had a more recent common ancestor than did, say, humans and gibbons, or humans and rhesus monkeys.

6 Macroevolution results in the divergence of taxonomic groups, in which the descendant is in a different taxonomic group to the ancestor. Table 12.2.1 illustrates this concept because it shows how a common ancestral primate has evolved into five different species.

12.3 CASE STUDY: WHAT DID TERROR BIRDS EAT?

1 *Gastornis* was assumed to be a predator because it looked very fierce and it had a huge, sharp beak to grab and break the neck of its prey. Two pieces of evidence that pointed away from this before the so-called 'geochemical approach' were that it did not show imprints of sharp claws, a feature needed to grapple prey and the was so large that it would not be a very swift hunter.

2 After the mass extinction of the dinosaurs, surviving mammals, birds and flowering plants underwent evolutionary radiation because major competitors were wiped out, and new species had unprecedented access to new habitats and no longer had to compete for space, food and water.

3 Missing raptor-like toe claws on *Gastornis* may have been an artefact, as they may have not made an impression in the soil when the footprints were made before fossilisation.

4 The approximate calcium isotope composition of omnivores would be somewhere between that of carnivores and herbivores. This is because the lighter isotopes of calcium become progressively enriched along a food chain and omnivores consume both plant and animal material.

5 If *Gastornis* was herbivorous it would have eaten tree ferns, the leaves and branches of large trees and possibly grasses and shrubs.

12.4 EVOLUTION OF THE MARSUPIALS

1 After continental drift broke up the southern continent of Gondwana, South America, Antarctica and Australia initially stayed together. This allowed the marsupial ancestor to move from South America to Australia via Antarctica.

2 The type of evolution shown in Figure 12.3.1 results in the divergence of taxonomic groups, in which the descendants are in a different taxonomic group to the ancestor. This falls into the definition of macroevolution.

3 A phylogenetic tree shows evolutionary relationships between related organisms and forks in the diagram are points at which lineages have diverged.

4 Time is represented by moving from left to right across the diagram.

5 Each fork on a branch marks a point at which new species arise. For example, consider the node at the fork where Macropus and Potorous separate. Various evolutionary events would have caused the populations to become so different that they could no longer interbreed. This means that each node represents an ancestor common to the species above that node.

6 The member of Dasyuromorphia that would have DNA most similar to the genus Dasyurus, is the genus Phascogale.

7 In the order Diprotodontia, Tarsipes, the honey possum, and Pseudocheirus, the ring tail possum are the most closely related as they have the most recent common ancestor.

8 This study could be called comparative genomics because comparative genomics uses a comparison of whole or large parts of genome sequences of different species to discover evolutionary relationships.

9 Large scale and exhaustive computations are important in a study of this kind because comparing the complete genome sequences of different species produces huge amounts of data that must be analysed by a computer program.

CHAPTER 12 EVALUATION

1 a The evidence used to determine the evolutionary relationships shown could have come from the comparative anatomy of body structures and fossilised structures, such as bones and teeth or from a comparison of their DNA.

b Major extinction events are followed by evolutionary radiation because with the destruction of so many species, major competitors were wiped out, and survivors have access to new habitats and no longer had to compete for food and water.

c The evolutionary relationships of Australian marsupials and monotremes, provide strong support for Darwin's concept of descent with modification because Figure 12.2 shows that marsupials today have descended and evolved from common ancestors, that were different to their modern descendants.

d The closest relatives of the extinct diprotodonts are wombats and koalas.

2 Both Lamarck and Darwin proposed a theory of evolution. Lamarck suggested that organisms pass on to their offspring characteristics that they acquire during their lifetimes. By contrast, Darwin's mechanism for evolution was natural selection; the process where individuals with certain inheritable traits survive and reproduce more successfully than other individuals, leading to evolutionary change in the population.

3 Microevolution is the outcome of natural selection, which is a change in the frequency of various alleles within a population. Although macroevolution also involves a change in the frequency of various alleles within a population, over geological time macroevolution results in the divergence of taxonomic groups, such that the descendant is in a different taxonomic group to the ancestor.

4 As the DNA code is universal in all living species, it links all life on Earth to a common ancestor and is strong evidence for evolution. Comparative genomics produces a detailed picture of DNA sequence conservation to reveal the shared common ancestry of diverse species which is also strong evidence for evolution.

CHAPTER 13 REVISION

13.1 GENE POOLS

1 A population is a group of individuals of the same species that live in the same geographic area and readily interbreed to produce fertile offspring.

2 A gene pool is the total collection of alleles present in a population.

3 Student answers may vary. An example would be C^B – brown, C^W – white

4 a Dark brown mouse C^BC^B

b Light brown mouse C^BC^W

c White mouse C^WC^W

5 320 alleles for coat colour.

6 a C^B alleles: 90 in light brown mice, 55 × 2 in dark brown mice = total of 200 in gene pool of 320 alleles.
Frequency = 200 ÷ 320 × 100 = 62.5%

b C^W alleles: 90 in light brown mice, 15 × 2 in white mice = total of 120 in gene pool of 320 alleles.
Frequency = 120 ÷ 320 × 100 = 37.5%

7 Light brown mice would have the greatest camouflage in the hay. White mice would be more visible against the hay colour compared to dark brown mice. Therefore mice with the brown allele have a greater chance of surviving and reproducing more offspring than those with the white allele alone.

8 New alleles arise from existing alleles that have mutated. Mutation is the only source of new variations within populations.

13.2 CHANGES TO ALLELE FREQUENCIES

1 B alleles: 12 alleles out of a gene pool of 24 alleles = 50%

b alleles: 12 alleles out of a gene pool of 24 alleles = 50%

2 a B alleles: 18 alleles out of a gene pool of 30 alleles = 60%

b alleles: 12 alleles out of a gene pool of 30 alleles = 40%

b B alleles: 12 alleles out of a gene pool of 18 alleles = 66%

b alleles: 6 alleles out of a gene pool of 18 alleles = 33%

3 In the mouse population when three dark mice migrate in: C^B alleles: 90 in light brown mice, 58 × 2 in dark brown mice = total of 206 in gene pool of 326 alleles. Frequency = 206 ÷ 326 × 100 = 63.2% compared to 62.5% in the original population.

In the mouse population when three white mice leave: C^W alleles: 90 in light brown mice, 12 × 2 in white mice = total of 114 in gene pool of 314 alleles. Frequency = 114 ÷ 314 × 100 = 36.3% compared to 37.5% in the original population.

The smaller the population size, the greater the impact of migration on allele frequencies.

9780170411745

4 a Genetic drift

Description: If a population is small, there is a chance that some alleles present in a parental group will not be passed on at all. These alleles may be permanently lost from the gene pool.

Graphic representation:

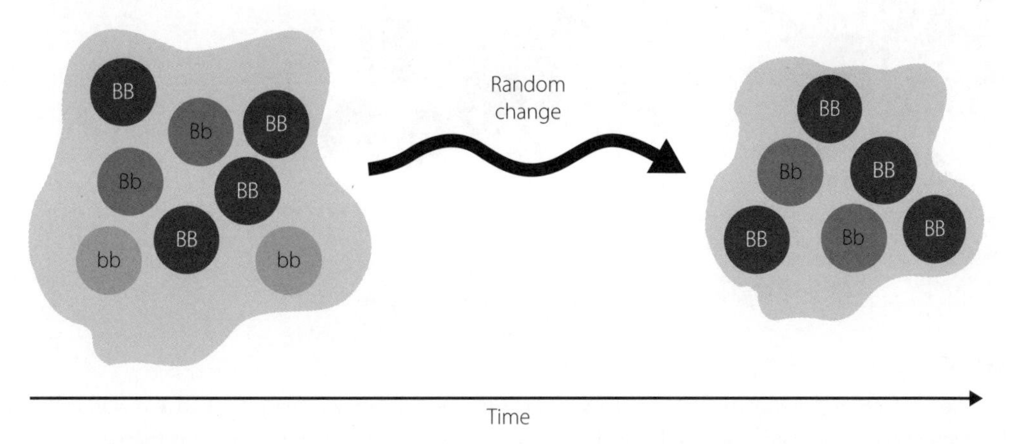

b Bottleneck effect

Description: A catastrophic event or a period of adverse conditions drastically reduces the size of a population. Certain alleles may be lost through chance.

Graphic representation:

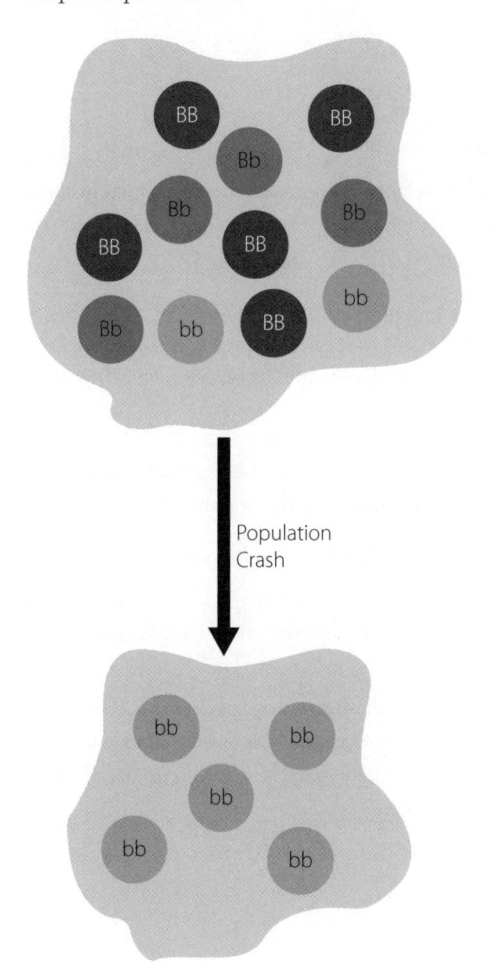

c Founder effect

Description: The founder effect occurs when a few individuals carry alleles to a new, isolated area and a new population is formed with different allele frequencies to the original population.

Graphic representation:

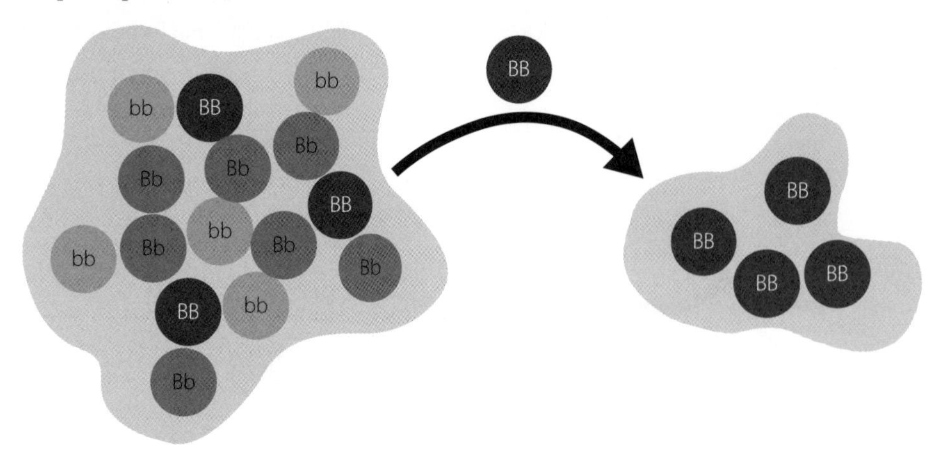

13.3 NATURAL SELECTION

1 Without genetic variation, selection pressures will act on all individuals in a population equally. All will survive and reproduce at the same rate. For natural selection to act, some individuals need to survive and reproduce better than others.

2 Step 1: Mutation creates variation.

Step 2: Unfavourable mutations are selected against.

Step 3: Reproduction and mutation occur.

Step 4: Favourable mutations are more likely to survive.

Step 5: Favourable mutations are more likely to reproduce.

3 Step 1: Mutation creates variation. *B. betularia* occurred in different varieties: white speckled colour or black colour.

Step 2: Unfavourable mutations are selected against. During the Industrial Revolution when trees became covered in soot, the white speckled coloured moth was selected against as the birds could see them more readily. The black coloured moth was selected for.

Step 3: Reproduction and mutation occur. The moths continued to reproduce and mutate at the natural rate.

Step 4: Individuals with favourable mutations are more likely to survive. As more black coloured moths were able to evade birds and survive, they reproduced in greater numbers.

Step 5: Individuals with favourable mutations are more likely to reproduce. As time went on more black coloured moths reproduced than white coloured moths; hence, there were greater numbers of black moths in the population.

4 a Stabilising selection: As long as the environment of an organism is not changing, the selective pressures will act against deleterious alleles that cause a departure from the optimal phenotype.

b Directional selection leads to a change in the frequency of a trait over time. Changes in environment lead to selective pressures favouring organisms with new or more extreme traits.

c Disruptive selection operates in favour of extremes. For example, a drought may kill off a local species of shrub that produces medium-sized seeds. A species of seed-eating bird may experience disruptive selection in this situation, when there are only large seeds (or only small seeds) available to eat. Birds with intermediate-sized bills would not be as well adapted for eating either the large or the small seeds and would be selected against.

5 a Disruptive selection: extreme traits are favoured

b Directional selection: the trait shifts in one direction

c Stabilising selection: the trait stabilises

6 a In this situation, b alleles will increase in frequency and B alleles will decrease in frequency.

The white mice are camouflaged better against the white background compared with brown mice. More white mice will survive predation and are able to produce more offspring than brown mice. As time goes on the population contains more white mice, increasing the allele frequency of b alleles.

b Directional selection. Coat colour allele frequency shifts in one direction.

CHAPTER 13 EVALUATION

1 a The frequency of occurrence or proportions of different alleles of a particular gene in the total collection of alleles within a population.

b 100 alleles in the gene pool.

c s allele in generation 1 = 50 out of a gene pool of 100 = 50%

s allele in generation 2 = 66 out of a gene pool of 100 = 66%

s allele in generation 1 = 86 out of a gene pool of 100 = 86%

d Natural selection is the likely process that caused the change in allele frequency. Selection pressures changed causing the fish with stripes to be less suited to the environment. Fish without stripes were at a selective advantage and survived better and reproduced more offspring.

e Directional selection is the term that is shown in this example. Changes in the environment lead to selective pressures favouring organisms with new or more extreme traits and leads to a change in the frequency a trait over time.

2 a This is most probably due to the founder effect. A small number of individuals have moved to this island and become isolated from the larger British population. By chance there was a higher frequency of retinitis pigmentosa alleles in the founding population compared to the British population.

b If the natural disaster caused a drastic reduction in size of the island population, the allele frequencies could be a result of a bottleneck effect. It may be that a portion of the population survived the disaster and by chance they carried a higher frequency of retinitis pigmentosa alleles.

3 Originally there was an inherited variation in resistance to the myxoma virus. A small number of resistant rabbits existed before the myxoma virus was introduced. After the myxoma virus was introduced more resistant rabbits survived and reproduced more offspring than non-resistant rabbits. During the 1950s the frequency of resistant rabbits increased.

4 Correct part of statement: Natural selection acts on traits that make some individuals better suited to their environment than others.

Incorrect part of statement: Individuals develop mutations.

'Mutations occur by chance and create variations of characteristics. Natural selection acts on individuals with certain traits that suit them better to their environment compared to other members of the population.'

CHAPTER 14 REVISION

14.1 MECHANISMS OF ISOLATION IN SPECIATION

1 Groups of actual, or potentially, interbreeding natural populations that are reproductively isolated from other such populations, are defined as a biological species.

2 Interbreeding patterns of the long-dead dinosaur are not possible, only the physical and physiological characteristics are able to be observed. The morphological species concept defines a species using measurable anatomical criteria and characteristics.

3 The South American and Australian continents started drifting apart around 180 million years ago. Dinosaur populations became physically separated and could no longer interbreed. Gene flow between the populations ceased. The gene pools accumulated changes over time. Selection pressures favoured different alleles in each of the locations. Eventually the changes became so great interbreeding would not have been possible, the populations would become reproductively isolated and new species formed.

4 Allele frequencies would remain much the same and new species would not form. Speciation depends on different selection pressures acting on the populations.

5 Pre-reproductive isolating mechanisms prevent organisms from being able to interact to reproduce while post-reproductive isolating mechanisms prevent fertilisation occurring or an embryo developing into viable offspring if fertilisation does occur.

6 a Geographic: individuals are separated by geographic features, such as seas, mountains, distance or habitat.

b Temporal: individuals breed during different seasons of the year or times of the day.

c Behavioural: individuals have different courtship patterns.

d Morphological: individuals have different reproductive structures so that mating is physically impossible, for example genitalia of different size, shape or location.

e Spatial: individuals are separated from each other by distance.

7 a Gamete mortality: the gametes do not survive.

b Zygote mortality: the zygote forms but does not survive.

c Hybrid sterility: adult offspring are formed but are infertile because they are unable to produce viable gametes.

8

EXAMPLE	TYPE OF ISOLATING MECHANISMS
A horse and donkey produce an infertile mule	Hybrid sterility
Variation in an animal species size prevents the largest in the species mating with the smallest	Morphological
A massive earth quake creates a mountain range separating populations from each other	Geographic
A population of snakes emerges from hibernation at a different time to the main population	Temporal
Coral forms in a cooler area of a reef	Spatial
Pollination and fertilisation of flowers is successful but no seeds develop	Zygote mortality
Plants flower at different times of the year	Temporal
The mating call pattern of a frog changes	Behavioural
A new river flow forms	Geographic
Reproductive structure shape of snails differs	Morphological
A species of rice pollinates another species of rice that produces seeds that grow into plants that are sterile	Hybrid sterility
Birds differ in their mating rituals	Behavioural
Gametes no longer fuse together	Gamete mortality

14.2 MODES OF SPECIATION

1 a Allopatric speciation takes place when organisms that could interbreed do not do so because their ranges do not overlap owing to geographic isolation. Geographic barriers physically prevent individuals of a species from associating to breed. Geographic isolating mechanisms include large bodies of water such as seas, mountain ranges and changes to habitat due to land clearing and desertification.

b Sympatric speciation refers to the evolution of two or more new species from a single population within the same place. It requires a reproductive barrier that isolates members of a population from the rest of the population in the same area. If gene flow between the isolated population and main population is prevented and different selection pressures act on the isolated populations, allele frequencies may become so different that individuals may be unable to interbreed, resulting in evolution of new species from a single population within the same place. Examples could include the case of the difference in hatching time in Magicicada cicadas and changes in mating calls of frogs.

c In parapatric speciation individuals are more likely to mate with individuals in their geographic area rather than individuals in a different area. Gene flow would still continue in the bordering areas but over time the populations would diverge to become better adapted to the different conditions in different areas of the environment. Examples could include grasses growing in areas of contaminated soil around the border of non-contaminated soil.

2 a Habitat fragmentation is the process by which areas of a habitat are lost, resulting in division of a large continuous habitat into smaller, more isolated habitats.

b Gene flow between populations in the isolated habitats is prevented. Over time subspecies will develop if selection pressures in the local areas become different and allele differences accumulate.

c Provide linking wildlife corridors (also known as habitat or green corridors) of natural landscape. These allow animals to move to new locations when resources become scarce, to facilitate seasonal migration and to permit interbreeding, ensuring that there is sufficient gene flow between different parts of the isolated populations.

9780170411745

3

	ALLOPATRIC SPECIATION	SYMPATRIC SPECIATION	PARAPATRIC SPECIATION
ORIGINAL POPULATION			
FIRST STEP OF SPECIATION			
REPRODUCTIVE ISOLATION			
NEW SPECIES			

14.3 PATTERNS OF EVOLUTION

1 a Divergent evolution is a process whereby related species evolve new adaptations over time, away from the common ancestor, to give rise to new species.

b Convergent evolution is a process whereby unrelated organisms evolve similar adaptations in response to similar environmental pressures.

c Parallel evolution is a process whereby unrelated organisms evolve similar adaptations in response to the same environmental pressures.

d Coevolution is a process whereby an evolutionary change in one species has affected an evolutionary change in another species.

2 a

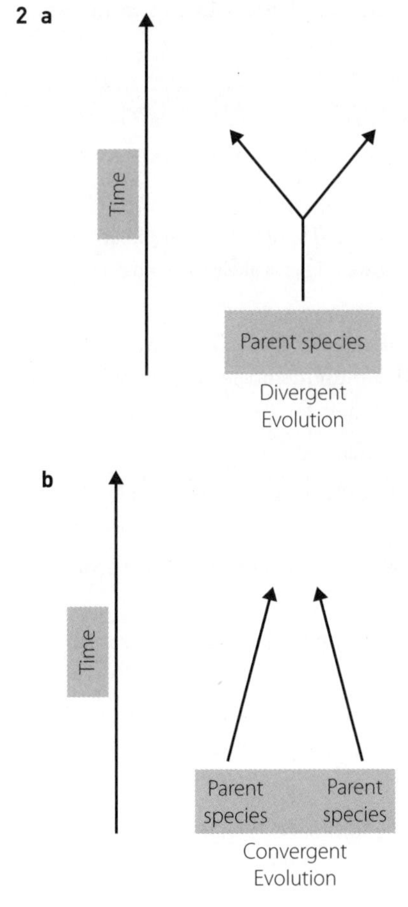

c

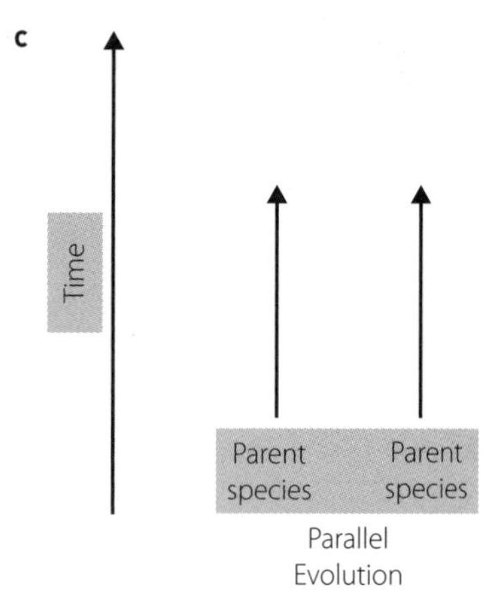

3 Both the dunnart and mouse have developed a similar body plan independently, rather than it being a legacy from their common ancestor. The results of this convergent evolution are adaptations that are similar to solve a problem in a similar way.

4 Divergent evolution causes this diversification. The Tasmanian devils and marsupial moles live in different types of habitats and feed on different types of animals. They have evolved different adaptations to suit them to their way of life.

5 **a** Coevolution.

b An example could be the evolution of flowering plants and their pollinators.

6 **a** Convergent evolution. Wing features and the flying ability of all three animals are similar however they have not developed from a common ancestor.

b Parallel evolution. The same degree of similarity has evolved over time.

c Divergent evolution. The mammals had a common ancestor but due to adaptations to different environmental pressures, they have quite different features.

7 When only a small number of individuals survive a population bottleneck event, the surviving population are unlikely to carry all the alleles that were present in the original population. This results in low genetic diversity. Inbreeding within a small population further reduces the gene pool. If this population is then exposed to a changed selection pressure, natural selection will act on 'fit' individuals with the best suited alleles for survival and reproduction. When genetic variation is low there is less chance of the presence of alleles suiting the selection pressure. If no individuals in the species have the right genetic variation present, the species will become extinct.

CHAPTER 14 EVALUATION

1 The group needed to show that spiders in each new species did not produce fertile offspring with members of known species.

2 **a** Divergent evolution.

b This suggestion is incorrect as a physical barrier is needed for allopatric speciation to occur. There is no physical barrier involved with this example of bat diversification. The different species live in the same area. This is likely to be an example of parapatric speciation.

3 Having a low genetic diversity means members of the species are all genetically very similar and so respond in a similar way. If there is an adverse selection pressure introduced such as a disease then it is likely that most members of the population are affected as there is very little variation.

4 Students can name one of: geographical, temporal, behavioural, morphological, and spatial.

5 Separated populations have different gene pools. Different selection pressures act on each population. If the two populations, when brought together, do not produce fertile offspring, they are different species.

6 **a** This diagram represents convergent evolution. Convergent evolution is a process whereby unrelated organisms evolve similar adaptations in response to similar environmental pressures.

b This diagram represents parallel evolution. Parallel evolution is a process whereby unrelated organisms evolve similar adaptations in response to the same environmental pressures.

7 **a** Coevolution is a process whereby an evolutionary change in one species has affected an evolutionary change in another species as shown in this example. Both the snake predator and the newt prey apply selection pressure.

b Student answers will vary. Pollination is good example.

8 New species form while geographically isolated in allopatric speciation whereas in sympatric speciation new species form while in the same geographical area.

In allopatric speciation gene flow ceases between populations because of a barrier. In sympatric speciation ongoing gene flow is possible.

PRACTICE EXAM ANSWERS

BIOLOGY UNITS 3 & 4

MULTIPLE CHOICE QUESTIONS

1 A

2 B

3 C

4 B

5 B

6 B

7 D

8 C

9 A

10 A

SHORT ANSWER QUESTIONS

1 a

	IDEA (1)	IDEA (2)
Mammals sharing the most features in common with the pig	Hippopotamus	Camel
Mammals sharing the least features in common with the pig	Whale	Wale, goat and hippopotamus

b Convergent evolution or analogous structures.

c Disagree because all organisms on Earth are related to one another, meaning that at one time they had a common ancestor.

d The more similar the molecular sequences (e.g. DNA and proteins) between groups, the more closely related they are regardless of the physical characteristics.

2 a

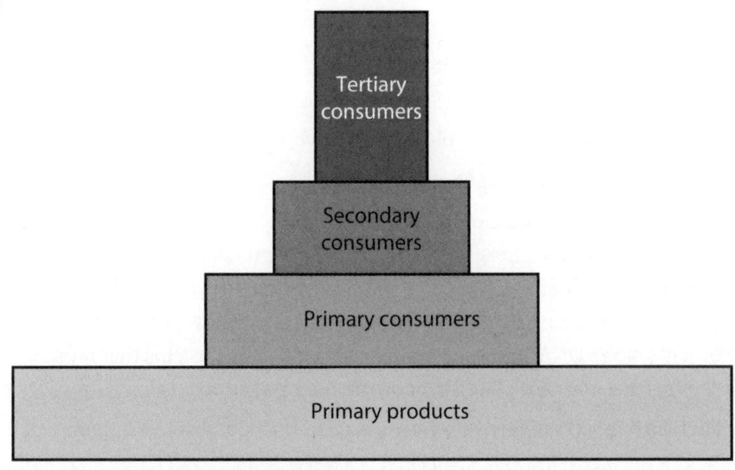

b Reason one: The remaining 90% of energy is transformed by metabolism into heat energy and lost to the surroundings.

Reason two: The remaining 90% of energy remains as chemical energy in both the uneaten portion of an organism and its body waste.

c Keystone species are organisms that provide essential services for a wide range of species in the ecosystem, in this case seed dispersal.

d By eating fruit and aiding with dispersal of seeds, both the bird and plant benefit (symbiosis). The birds can eat insects and mammals (predation).

e There would be fewer primary producers as seeds are not being spread. This would have an impact on higher trophic levels.

f Population growth rate = (birth rate + immigration rate) – (death rate + emigration rate)

$= (20 + 2) - (4 + 0) = 18$ per 80 = an increase of 22.5%

3 a i .....AGCTACGGGTGG..... The base change is from G in a normal allele to A in a mutant allele.

ii .....TCGATCCCACC.....

b A substitution mutation does not affect the sequence of other codons in the gene as there are still the same number of base pairs.

c In this case, a frameshift mutation has occurred. The starting point for the reading of the triplet code sequences for the amino acids to be included in a polypeptide chain is shifted away from the original position. As a result, all the codons 'downstream' of the mutation are affected. In the polypeptide produced with a frameshift mutation, the amino acid sequence beyond the location of the mutation bears no resemblance to those in the polypeptide that would have originally been produced.

d Both parents must be heterozygous. If the chromosomes with the normal allele from both parents are passed on to the offspring, the child is normal. If A represents the mutant allele and a represents the normal allele, the parental cross is shown as Aa x Aa. Possible offspring and their ratios for genotypes are 25% AA, 50% Aa, 25% aa. Possible offspring and their ratios for phenotypes are 75% with Achondroplasia and 25% normal.

e

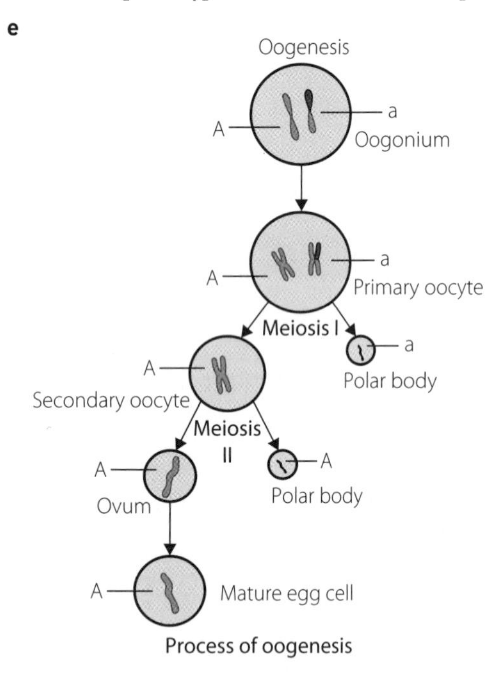

Process of oogenesis

f Crossing over increases the variation in the combination of alleles during meiosis.

g The enzyme helicase unwinds the DNA, separating the strands.

Free nucleotides attach to exposed nucleotides with the help of DNA polymerase enzymes.

DNA ligase enzymes seal the new nucleotides into a continuous strand.

h Regulatory proteins are products of genes that regulate the expression of genes other than their own. Specific regulatory proteins that bind to the DNA are called transcription factors. Those that activate gene expression bind to a section of non-coding DNA and enable it to unwind from histone proteins. Genes near this unwound DNA can then be transcribed. Activators also assist the binding of RNA polymerase to coding DNA segments to begin their transcription to mRNA. They may bind to several different specialised segments of the DNA, and even to introns. Regulatory proteins that repress gene expression may bind to a particular region on the DNA and block the RNA polymerase from attaching for transcription.

4 a Diversity among the insects is likely to remain constant if the selective pressure is equally favourable to all phenotypes.

b Gene flow between the mainland population and the population on Lord Howe Island was disrupted when populations became physically separated through geographic isolation. There were different selection pressures on the populations. As genetic changes accumulated the populations could no longer interbreed and became different species.

c The founder effect is a particular example of gene flow. A small number of stick insects may have drifted onto Lord Howe Island and become isolated from the larger mainland population. The island stick insects might not have carried all the alleles that were present in the mainland population. Over time changes to the genetic pool brought about by different selective factors meant interbreeding could no longer occur and speciation occurred.

d This is an example of convergent evolution. Unrelated organisms evolve similar adaptations in response to their environment when selection pressures are similar.

e There was very little genetic diversity in the population. When Black Rats arrived on the island, a new selective pressure was introduced. Most members of the population were at a disadvantage and this led to a drastic reduction in numbers.

9780170411745